…plate …enger

A particular witness to the Steam Age

The *Locomotive Journal* writing of E.S.Waterhouse

Compiled & edited by Robert Waterhouse

Foreword by Paul Salveson

Photography by Denis Thorpe

Baquis Press
Manchester

Dedication

In memory of my brothers David (see below and page 76) and John, fellow engine-spotter in the early 1950s on the King's Cross mainline near Alexandra Palace

Acknowledgments

Resurrecting material published decades ago is a bittersweet experience, especially material created by a grandfather. In the family we all knew about ESW's attraction to steam, but perhaps strangely he didn't share much of it with us. My brothers and I heard more about his other secular passions, debated during country walks with his cocker spaniel Chum or while competing on the council putting green.

So I'm grateful to my late brother David for collecting remnants of ESW's busy life in the family archive, including those relating to the *Locomotive Journal* columns. To read bound volumes of these magazines today one must join the British Library; they are also preserved by ASLEF, where Keith Richmond and Christine Farley were very helpful.

I spread draft copies of the introduction and sample columns around friends, all of whom responded usefully. Richard Challoner, Chris Dodd, Bill Hall and Chris Johnson were particularly perceptive in their comments. Then, for expert advice, I was introduced by Kevin Gopal to Paul Salveson. I'd like to thank him for his vital suggestions, his good-humoured patience and for agreeing to write the foreword.

Two small books by Ernest F. Carter have proved an invaluable reference for checking matters of detail: *Railway Locomotives of Britain*, in Frederick Warne's The Observer's Book series, London, 1955; *British Steam Locomotives*, Foyles Handbooks, London, 1962.

The columns were not illustrated as such. My grandfather left behind many standard photos of engines, but I turned to my former colleague and longtime friend Denis Thorpe for more interesting images, previously unpublished.

Alan Ward set his design skills in new directions while Nick Jaspan kept Baquis Press's Manchester base in business. Monique Grazzini, her family and team of Hotel Mas de Vence offered a delightful alternative work environment.

My wife Jane lived in harmony with yet another project. I'd like to thank her for all her encouragement and support over the years.

Robert Waterhouse
Manchester March 2024

Colophon

ISBN 978-1-7398873-2-2

Cover:
Great Eastern B12 on mainline service near Northampton

Opposite:
E.S.Waterhouse on the footplate of GWR 6001, King Edward VII

6001

Contents

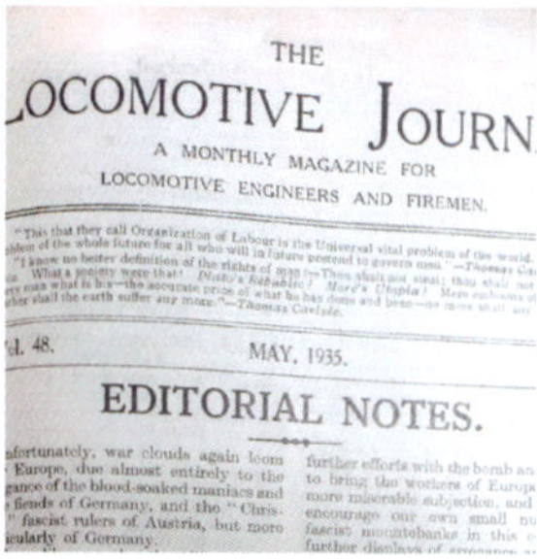
THE
OCOMOTIVE JOURN
A MONTHLY MAGAZINE FOR
LOCOMOTIVE ENGINEERS AND FIREMEN.
MAY, 1935.
EDITORIAL NOTES.

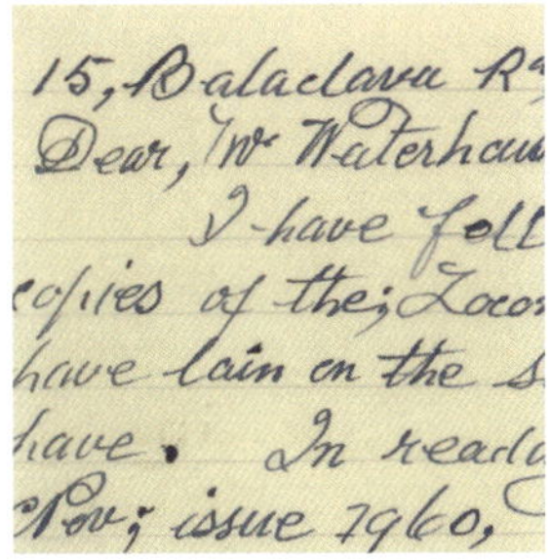

15, Balaclava R
Dear, Mr Waterhou
I have foll
copies of the Locos
have lain on the
have. In read
Nov; issue 1960,

Foreword:
Paul Salveson

It isn't often you can say that a new book on railway history is truly ground-breaking. This one is. The writings of E.S.Waterhouse are little known and his grandson Robert Waterhouse has done a great service in making a selection of them available to modern readers.

Why are they so special? Railway historians are familiar with the writings of E.L.Ahrons, and more recently Cecil J. Allen and O.S.Nock, who recorded locomotive performance across Britain. They were written from the perspective of well-informed middle class 'outsiders'. While they had a degree of sympathy for the men who worked on the footplate, it only extended so far and they were generally hostile to trades unionism.

Waterhouse was different. He was certainly from an educated middle class background but was very much in sympathy with the men whose work he described so well. It was remarkable that his writings were published in the locomotivemen's magazine, the ASLEF *Locomotive Journal*. His column, "As the Passenger sees it", ran for several decades and it is clear that Waterhouse was treated as part of the community of enginemen.

Waterhouse must have become a familiar face amongst locomen at the major London depots such as Old Oak Common, Camden and Stewart's Lane. I wonder what they thought of him? Perhaps some were slightly wary of this college professor and gentleman of the cloth, sometimes possessing a footplate pass, writing about their work. However, the decisive difference between Waterhouse and, say, Cecil J. Allen, was that he was writing not for the anti-union *Railway Magazine* but for their own monthly journal. His notes, as he called them, would have been read in the messrooms of Nine Elms, Norwich, Crewe North, Longsight, Laira and Polmadie. That's a great accolade for any railway writer, and the fact that his column continued for so long is proof that he was very well liked. Readers' letters show mutual trust and understanding.

In this cynical modern world some of his writing might seem slightly archaic, describing particular drivers as 'Brother'. I find that touching and shows the depth of his own admiration for the railway fraternity of the day.

And it was, of course, very much a *male* fraternity. It is only relatively recently that women have become drivers in this country. I suspect that Waterhouse would have welcomed this development; he was a man of what was termed 'advanced views'.

Waterhouse was very aware of developments in locomotive practice and performance across the world. The record here of a 1936 visit to the United States and Canada makes for fascinating reading, not least his account of trades unionism on the American railroads. He is unlikely to have met Eugene Debs, locomotive engineer and socialist presidential candidate – but I think they would have got on famously!

The writings reproduced here are of great value for a number of reasons. First, they shed light on working conditions on Britain's railways, particularly on the footplate, during a time of great change. His columns take us through the 1930s and the Second World War, then well into the period of nationalisation.

Waterhouse makes clear his views on modern traction, which were not sympathetic! He was very much a steam man and bemoaned the early extinction of steam on British Railways. Perhaps many of his friends on the footplate didn't share his views and welcomed the transition to cleaner, less arduous work on diesel and electric traction. This was gently pointed out to him by the Scottish engine driver and chronicler of ASLEF history, Norman McKillop (aka Toram Beg), with whom he struck up a great rapport.

The second area of value of these writings is more specialised. Within the rail enthusiast community there is continuing interest in locomotive performance, particularly that of steam, across the twentieth century. Waterhouse's detailed accounts of 'runs' on Britain's mainline railways gives a lot of new material to the 'performance expert'. His thrilling description of runs on the footplate of Castles, A3s and Royal Scots includes detailed accounts of regulator and valve settings which I certainly found interesting, and I'm sure some others will too.

Perhaps another aspect of their value is the light shed on English eccentricity. The fact that a Methodist minister became privy to the very specialised world of steam locomotive operation is, in itself, very special.

Waterhouse was not a typical railway passenger and the title of his column has always struck me as slightly ironic. The vast majority of passengers had little idea of conditions on the footplate, the demanding nature of the work, but also the intense pride in the locomotiveman's craft.

Up to now, the railway writings of E.S.Waterhouse had mainly been seen by locomotivemen themselves; few will be alive who remember the original column in their *Journal*. Bringing his work to a wider readership, nearly a century after his first column appeared, is very welcome. This is an important contribution to the social history of our railways.

Introduction: **Robert Waterhouse**

Between May 1935 and early 1964 my grandfather, Eric Strickland Waterhouse, wrote a column for the monthly *Locomotive Journal*, ASLEF's house publication, under the general heading "As the Passenger sees it" and the byline E.S.Waterhouse. His column noted specific rail journeys, who handled them, how man and machine performed against timetable and the challenges on each occasion. It also discussed issues facing enginemen. In short, the column put a human face to a craft often carried out in adverse conditions against increasingly demanding schedules.

Writing to ESW (as I will call him) in December 1935, John Bromley, ASLEF general secretary and de facto editor of the *Journal*, said "Your articles are causing quite a lively interest amongst our members, as it is something new to them to have the names of the enginemen given and with such kindly reference to their work." Bromley, general secretary from 1914, a Labour MP for Barrow-in-Furness between 1924-1931, retired in 1936. He was the first of seven editors for whom ESW was to contribute a column most months until shortly before his death.

Just who was 'The Passenger' and what right did he have to discuss railway issues with professionals? My grandfather, born 1879 in rural Leicestershire, was the son of a local doctor who died aged only 29 in a diptheria epidemic he tended. The young Eric and his brother Carl were taken by their mother to live in Bangor, North Wales. Eric won a scholarship to attend Rydal School, Colwyn Bay, as a day pupil, and it was before the morning rail journey from Bangor, as he describes in his very first *Journal* column (page 16), that he fell in love with steam locomotives and those who drove them.

It wasn't simply a schoolboy crush. We learn from a much later column (page 125) that, sometime in the mid-1890s, ESW was offered an apprenticeship at the Nine Elms works of the London & South Western Railway by no less a figure than Dugald Drummond, designer of the celebrated Drummond 0-4-4 tank engine and the company's locomotive engineer. Yet ESW, to use his own words, was "not able to take his offer."

Family lore has it that his deeply-religious mother was set on his studying for the Methodist ministry. And so it happened. He enrolled, aged just 19, at Richmond College, Surrey, a Wesleyan Methodist training institution, was ordained in 1901 and found circuit work in South London, where during the First World War he also served as a hospital chaplain. Continuing his studies extramurally, he took a doctorate from London University, rejoining Richmond College in 1920 as a lecturer, then as vice-principal and eventually principal. The author of a dozen teaching books, he was an early BBC radio broadcaster and, as professor of the philosophy of religion from 1931, served on the Senate of London University. By 1935 and his first *Journal* columns he was in demand around the country – lecturing, examining and of course preaching.

It's doubtful whether ESW would have achieved similar success as a locomotive engineer. He was a near contemporary of Sir William Stanier and Sir Nigel Gresley. He had the luck to be born when steam locomotive design would reach its apogee, but no family background in business or engineering. However Drummond, notoriously cantankerous, had spotted talent and ESW's writing shows thorough understanding of steam technology. He had ambition and application. How far might he have gone?

What's clear is that ESW had the ability to mingle passions without one impacting on the other. The many hundreds of rail journeys he annotated, among them those spelt out in his columns, were largely work trips. He may have been on his way to examine in Sheffield, lecture in Norwich or join a Plymouth committee, but once on a train he was wholly 'The Passenger'. He would travel on the footplate if he had time to secure a permit, but most of his trips were made in the front coach, where he was best placed to speak to drivers and firemen before and after journeys. It was also a likely spot to meet ASLEF members travelling between jobs.

There's no suggestion that ESW had any interest in proselytising. One or two letters in his small archive come from Society members who were also Methodists – not surprising given non-conformist union ties. ESW would have been wearing his dog collar if going straight to a meeting. That must have caused discussion. Driver C.Banks of Cricklewood wrote to him in March 1948 wondering, amongst other matters, why so many clergymen were interested in railway engines. "I've often tried to puzzle this out. The only conclusion I can come to is that engines and trains are a good example of the blessings of the straight and narrow path, and they illustrate very convincingly what happens when they leave it."

We don't know how, or if, ESW replied to this. He tended to correspond by postcard and didn't keep copies. Nor do we know whether he met the Rev Wilbert Audry, author of the Thomas the Tank Engine stories, the celebrated engine photographer Bishop Eric Treacy, or the gloriously eccentric Rev E.R.'Teddy' Boston, who constructed the Cadeby Light Railway around his rectory and drove a vintage steamroller he'd acquired from Leicestershire County Council for parish visits. I doubt it.

Another reader, Fireman Noel Turner of Norwich, wrote in January 1946 with a different query: "A question overheard in the messroom aroused a great deal of interest. 'Does Mr Waterhouse ride on the footplate or on the train?' One said the heading of the article supplies the answer (passengers do not in general ride on the footplate). I cannot think of anyone better to ask than yourself, so I am taking this opportunity." Let's hope he received a suitable reply.

In his columns ESW never specifically defined his role. He had passenger interests at heart in the sense that he sought services that were reliable, on time and moderately priced. He mentioned comfort only rarely and was not particularly interested in speed for its own sake. He wrote principally about mainline passenger services and was occasionally taken to task by Society members working goods and local routes for neglecting them.

During 1940 Norman McKillop, an engine driver based at Edinburgh Haymarket, began a *Journal* column headed Scottish Opinion. McKillop, born in the Gaelic-speaking Highlands, wrote under the pen name of Toram Beg. He was a one off, driving a Gresley A4 Pacific on the East Coast Main Line but also a trained and accomplished writer. ASLEF commissioned him as author of the history of the Society, *The Lighted Flame*, published by Thomas Nelson in 1950 to mark its seventieth anniversary. Toram Beg was increasingly used by *Journal* editors to discuss strategic issues, not just in Scotland, while the space ESW filled was roughly halved.

This didn't lead to animosity between the two. Far from it. When, early in 1946, ESW wrote to McKillop asking to be put on the mailing list of the *Locomotive Express*, started in Edinburgh to reflect Scottish and technical matters, McKillop replied: "We consider E.S.Waterhouse if not one of ourselves at least very near to that. Your contributions to the *Journal* prove that you enter into the very spirit of all our problems with sympathetic understanding."

For there were problems. During the mid-1930s the British railway network was reaching a performance peak spurred by competition between the four rail companies formed under the 1921 Railways Act, the resulting high-speed passenger locomotives of C.B.Collett. R.E.L.Maunsell, Stanier and Gresley, premium services between London and major cities and a number of world firsts for steam traction. However, pay and working conditions had not followed suit. In the Depression of 1929, unions had agreed to a temporary 2½ per cent reduction of their historically modest wages which left them struggling throughout the 1930s, unrewarded for the increased responsibilities of bigger locomotives and faster trains, while the economy rebounded.

An ASLEF strike in August 1939 was averted only by the outbreak of war. When the four railway companies came under national control shareholders benefited from the generous subsidies believed necessary to keep the nation's backbone working during hostilities. For their part, however, railway workers found themselves on the front line of the Home Front, targeted by enemy action, their lives disrupted by multiple demands while inflation shrank wage packets.

By the war's end, ESW noted, slower timetables were harder to keep than prewar high-speed schedules had been because of depleted rolling stock and degraded track infrastructure. Meanwhile ASLEF fought continuing skirmishes for better pay and working conditions amid the austerity of a victory in which the railways had played a crucial role but were never to recover their importance in the national transport framework.

As a contributor to the *Journal*, ESW soon made it clear that, although not a trades unionist himself, he was at one with the objectives of organised labour. But politics was the domain of the general secretary, who began his Editorial Notes fronting the May 1935 issue: "Unfortunately, war clouds again loom over Europe, due almost entirely to the arrogance of the blood-soaked maniacs and dope fiends of Germany..." Downpage, focusing on the domestic scene, he wrote: "The National Government has just brought forth its 1935 budget. As usual, capitalism takes more than it gives and the workers 'pay the piper'..." ESW's first column, headed "Some boys of the Old Brigade", was hardly in the same vein.

In it he sought to establish his credentials for writing about the craft by which his readership earned their living. Following schoolboy initiation on the Bangor-Colwyn Bay line, he recalled his friendship with drivers on the old North London line, competition between 'singles' and 'couples' on the Midland route out of St Pancras, the feats and eccentricities of enginemen like Peter Clow of Crewe and the stories they bred.

One note of controversy came in the final paragraph – a lament on modern pooling policies under which drivers were expected to make the best of whatever engine was on the roster. "What these old drivers who had their own engines and loved them would have said beats my imagination." It was a subject to which ESW would return – along with poor coal, interminable permanent way hold-ups or slacks, inexplicably long station stops, schedules which placed impossible demands on drivers and firemen, unfair practices like all-or-nothing eyesight tests, and careless management. His unchanging voice was: given the correct equipment and conditions, Society members will more than fulfil their duties.

This he substantiated with precise details for each journey made of personnel, engine, load, time between stops or places, reasons for not meeting schedule, overall performance. He'd developed this practice as a young man in the early 1900s, and taken to carrying two fob watches – one in each waistcoat pocket – for timekeeping backup. A fairly manic "passenger", but such notes gave him the wherewithal to discuss whatever issues arose. When, for instance, in March 1936 (page 30), weighing the pluses and minuses of the experimental Stanier Turbomotive, he could make as technical a judgement as the writer O.S.Nock or indeed Toram Beg. Just how his knowledge was acquired is a mystery – his archive offers no clues.

What spurred me to search out his columns in bound volumes of the *Locomotive Journal* was discovering in this small archive letters from drivers and firemen taking up specific points, asking questions or simply describing their experiences. These, together with correspondence from the railway authorities about requests for footplate permits (and their comments about the article drafts they requested), plus the columns themselves, form a backdrop to decades of industry evolution.

In making a selection from the 300 or so columns ESW contributed between 1935-64 I've found myself concentrating on the mid-1930s, the wartime years, then nationalisation through to the early 1960s when steam was approaching its end. I've interspersed columns with letters, some reproduced in facsimile, and the occasion when ESW wrote to the *Daily Telegraph* in defence of train drivers.

I've added brief comments as context at the end of many articles, and have adapted some original headlines. I've removed full points from capital letters denoting railway companies or engine types, and italics or quotation marks from names of engines, engine classes and train services. I've retained ESW's practice of sometimes mentioning engine names as well as numbers, sometimes just numbers. I've deleted any number of commas, and have by and large resisted the temptation to edit perhaps old fashioned turns of phrase.

As a visual appendix I'm showcasing on pages 135-144 images mainly taken in the Midlands during the 1950s by Denis Thorpe for himself, way before his celebrated rail heritage pictures in the *Guardian*. They are revealing about the state of postwar Britain and its railway infrastructure.

You'll find a steam glossary at the back of the book. This seems necessary because ESW was, after all, writing for railway professionals and felt no need to explain technical or slang terms. The glossary doesn't pretend to be inclusive. I would suggest that persistent exploration of Wikipedia and other internet sources will satisfy just about every query.

In a brief Endword I've attempted to summarise my grandfather's place in the story of steam. As a one-time engine-spotter I understand why people over the years have fallen under the spell of steam locomotion. Yet few non-railwaymen had the opportunity to observe it at such close quarters. Not many wrote as clearly as ESW. Perhaps nobody else enjoyed such a warm relationship with enginemen. That was his privilege and is his legacy.

For our purposes, ESW's story begins in North Wales. It ends near there. In March 1964, at the age of 84, he made a final work trip by rail from his home in Epsom, Surrey, to Colwyn Bay and Penrhos College, the sister school to Rydal where he served as a governor. The following morning, a Saturday, he came on by train to Manchester. I met him at Exchange Station and drove him to Hayfield, the Derbyshire Peak District town where our branch of Waterhouses comes from and where he by tradition conducted the Methodist Sunday School Anniversary service. He was in good spirits but nursing a bad cold. He died just a few weeks later.

Albert Griffiths, the Society's general secretary at the time, devoted his Viewpoint column in the June 1964 issue of the *Journal* to E.S.Waterhouse. "Locomotivemen have lost a good friend, our craft a great admirer, and our Movement a wholehearted supporter" he wrote. My wish, 60 years on, is that something of this empathy infuses continuing debates about the rail industry's future.

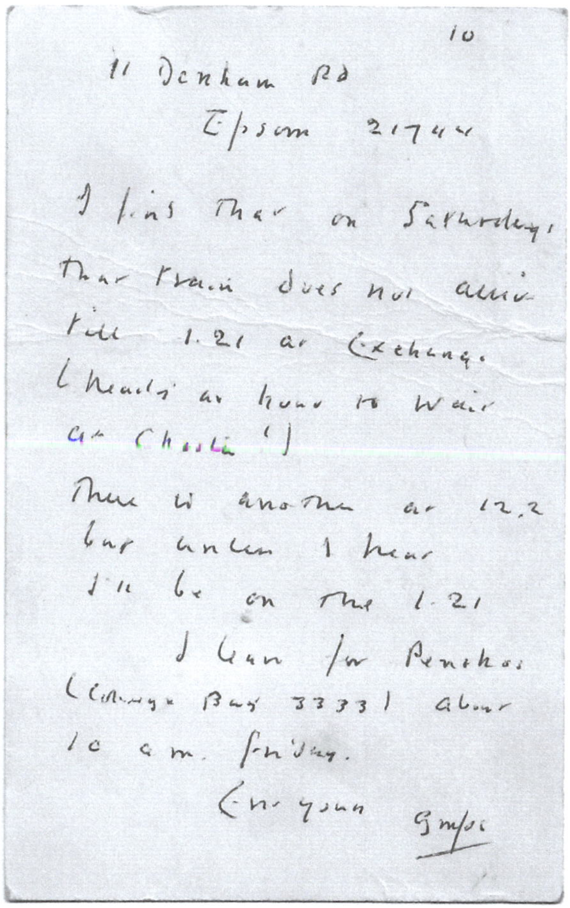

10

11 Denham Rd
Epsom 21744

I find that on Saturday, that train does not arrive till 1.21 at Exchange. (Nearly an hour to wait at Chester!)

There is another at 12.2 but unless I hear I'll be on the 1.21

I leave for Penrhos (Colwyn Bay 33331) about 10 a.m. Friday.

Ever yours

ESW's postcard advising his scheduled arrival time at Manchester Exchange Station, March 14 1964

1935-1937

'Time recovery is a very relative performance. A man with a light load can work marvels, where some unlucky fellow with a bad steamer and a tender full of rubbish may do magnificently even if he loses time'

June 1935

'Some of my old friends who once worked steam engines tell me they are getting used to electric traction and even like it, but I can't share their enthusiasm. I am with a driver of the old school who, asked if he would go on the electrics, said "No, I'd rather go back to goods; I am not a tram driver"'

December 1935

'It is no use grousing about faster schedules. Railway transport is fighting for its life against motor competition, and faster passenger and freight traffic is the only way of competing with road vehicles'

November 1936

'Were I an engineman I would be a unionist – not because, necessarily, I agree with all that the union officials think or do, but because I want to stand by my mates and to help for the common good of the whole'

October 1937

Some boys of the Old Brigade

May 1935

My interest in engines and enginemen began when, as a schoolboy at the end of the last century, I went by the 7.55 each morning from Bangor to school at Colwyn Bay. I was always at the station early, and one of the little Jumbos, Sirius 424 or Centaur 773, driven by Harry Williams or Owen Jones, would back on, and I would get on the footplate, warm my toes, and hear the wisdom of these men who were my heroes, until time for starting.

Later I came to London and lived on the old North London line. A driver had his own engine in those days. If it was No 4 you expected to see Brown, No 35 would be driven by "Tommy" Wingrove, No 48 by Hancock, a particular chum of mine. Many a ride I had on the footplate of these old 4-4-0 tanks, now all scrapped.

The alternate trains were worked by LNW "eight-wheel choppers", a class of engine none too good for their work, but drivers like Clarke, who had 786, performed wonders with them. My longer trips were mostly to Sheffield. I was a great admirer of the Midland singles, and though in those days loads were small compared with modern loading, for the power the singles did remarkable work.

One amusing incident is worth mentioning as an example. The 2pm from St Pancras to Leicester stopped at Kettering, 72 miles away, from 3.22pm to 3.26pm. The 2.10pm to Nottingham passed through without stopping at 3.24pm, using the other line, by the points cross-over south of the station. On the former train was one of Mr Johnson's new 4-4-0 engines of the 2607 type, standing on Platform 3 at St Pancras, whilst the 2.10pm, by which I was travelling, was on the opposite platform, and headed by a single, 1860, driven by Richardson of Kentish Town. The two drivers engaged in argument upon the merits of the couples versus the single wheel engine, Richardson stoutly maintaining against his colleague the case for the latter.

When we pulled out it was evident that Richardson was up to something. On the rising grade to Luton we made fair progress with a biggish load, but once Luton was passed 1860 began to fly. Bedford was reached five minutes early, and with a long shriek of the whistle we passed Kettering eight minutes before schedule, whilst the coupled engine was standing at the platform!

On another occasion the up Scotch express was eight minutes late at Leicester. I was particularly anxious to be in London on time as I had a connection to catch, and asked the driver, Barlow of Kentish Town, if he could recover a bit. "Wait and see" was the reply. The engine was also a single, No 130, and my hopes fell when at Wigston, a few miles from the start, signals brought us almost to a stop. This made us 11 minutes late past Market Harborough with 82¾ miles to go. Through the darkness and a snowstorm we raced and reached St Pancras dead on time, in 105 minutes net for the 99 miles. There are several trains today timed at that speed, but none even now are quicker, and for a single, with a load of nearly 250 tons, it represented great work.

Other trips were made on the old LNW. Many Crewe men recollect the name and fame of Peter Clow who drove Adriatic 1309 during the race of 1895. His usual engine was a compound of the same class, Coptic 1307, and Peter was of the opinion that he could take any load, disdaining the help of a "hooker". Sometimes the trip was too much even for his skill, for the compounds were bad starters and slipped a lot. On such occasions Peter would have furious arguments with the guard, demanding for signal checks and pwr delays more allowance than they deemed fit to cover time otherwise lost. One of the firemen, now a driver, told me that Clow would point to the firebox door and say "fire", and after a few shovels had been put on would cry "stop", and woe betide him if he dared to fire except under such instructions. But Clow was a tip-top driver and a great personality, and his little eccentricities were accepted as the accompaniment of genius.

Another "character" was an Irishman, also of Crewe, who once came to the shed and found himself down for the compound Stork 1397, notoriously a bad starter. "Stork, is it?" he snorted, "then it's meself that's stalking" and away he went. On another occasion he was hauled up for losing time, and blamed the coal. "Nonsense" was the reply, "the Company won't buy bad coal." "Sure, thin," came the ready retort, "they must get the divvle of a lot given them." They were great days, and great boys too.

The modern pooling of engines has economical advantages, but no others. What these old drivers who had their own engines and loved them would have said beats my imagination. Why, on the old Brighton line the driver's name was actually painted inside the cab. You may still see the name of William Love inside the cab of Gladstone in York museum. But what would they have said? Perhaps I had better not try to guess, or the Editor, who is careful about the language used by his correspondents, would censor me!

EDITOR'S FOOTNOTE

For his first Journal *column ESW establishes his knowledge and love of engines and enginemen back into the previous century, his ease on the footplate and his turn of phrase as a storyteller.*

Making up time is no competition

June 1935

So there is to be a new push about making up time. Certainly very few trains would make a punctual arrival if drivers made up no time. How seldom is there any long run clear of "sigs" and pwr? I have not calculated exactly, but I imagine that for every minute I book against an engine, I book 20 recovered. Indeed, it is the rarest thing for me to find an engine losing time unless grossly overloaded, or provided with bad coal. Occasionally one finds a driver whose methods recall those of the old "coal-dodgers" in the days of coal premiums, but that is very rare indeed. The vast majority keep and recover time when they can. But time recovery is deceptive.

The 10.25am from St Pancras to Leicester gets 109 minutes allowed on five days a week, and 119 on Saturdays. If, owing to late start, a driver recovered ten minutes on a Saturday, and was unlucky enough to be involved in an accident through no fault of his own, as likely as not the public Press would howl in headlines about "Terrible Smash! Train Making up Time is Wrecked" and so on. Yet obviously the train would merely be doing what on other days it regularly does.

If the companies want men to make up time, it is but fair that they guarantee them protection against such victimisation. It is also obvious that they must not grumble at a bigger coal consumption. Time recovery is a very relative performance. A man with a light load can work marvels, where some unlucky fellow with a bad steamer and a tender full of rubbish may do magnificently even if he loses time.

One of the best bits of loco work I have noted recently was when a driver with a Stanier taper-boiler 3-cylinder Jubilee had 439 tons behind and a firebox full of ash and clinker, owing to the fact that she had been sent out the moment she came in from a long trip because of the failure of the train engine. Crewe was left 14 minutes late, and we were the same time late into Euston, but it was splendid work under the circumstances, and full credit is due for it to Driver McKernan of Edge Hill Liverpool and his mate.

On the 1.20pm King's Cross to Newcastle, load 484 tons, Driver Samways recovered nine minutes' incidental delays to make right time at Newcastle. On the up train due at 9.55am, Driver Gutteridge recovered three minutes with no less than 540 tons, Peterborough to London, start to stop, 76¼ miles in 79 minutes, right time in.

These are not in any way special. They are simply taken at random from journeys I have made during the past few months, and many similar performances take place constantly. They are mentioned to show that drivers usually do recover lost time. But is it wise to make a kind of competition in time recovery? Would it not be better to tell drivers that they are to do their best and use their discretion?

The modern tendency is to start a train with a tight schedule and a lightish load. Bit by bit the load goes up, and at last it is the most difficult job to keep time. Doncaster men who work the 9.40am up from Grantham find that the loads have grown so that on many occasions over 400 tons is on, and the 105½ miles to King's Cross in 100 minutes wants some doing with 400 tons. Recovery of time is out of the question, though with a normal load I have known four or five minutes to be recovered on that schedule.

Would it not be better, since the public care far more about punctual arrival than for super-speeds, to allow a certain margin for time recovery on every schedule and encourage drivers to run to schedule, at the same time indicating what improvement on schedule was allowed for, in case of delays?

No one knows better than enginemen what is safe to recover. None knows better, too, how some of the distances from the distant to the home signal afford inadequate length for pulling up because the signal cannot be sighted sufficiently ahead, and many drivers nowadays shut off on approaching certain distants, purely because they fear that with the new tight schedules requiring extra speed they will be unable to pull up in time if the signal is adverse.

In a word, if time is to be recovered, trust the drivers, but give them a fair chance and a square deal on the job.

EDITOR'S FOOTNOTE

In his second column, ESW sketches out a theme he explores on later occasions – that enginemen recover time lost to delays if at all possible, when it's safe to do so. *For more on coal-dodgers, see page 70.*

LMS meets load and schedule targets

July 1935

In view of the way loads have increased and schedules have been cut in the past few years, the locomotive work of the LMS, and especially of the western section, has been highly creditable, and on my own trips on express trains I have found punctuality very good indeed.

I have only once travelled behind one of the new 6200 Princess Royal Pacifics, and in this case the load was too light to show what could be done over the Shap and Beattock summits, but the Royal Scots have responded well to the increased demands. I travelled on the 6.12pm up from Crewe on the first day of the new schedule that cut the time for the 152½ miles to Willesden to 142 minutes, an average of 64½ mph. Driver Farrell of Liverpool, with a train of 282 tons tare and 6140 of the Royal Scot class, cut the timing by 5½ minutes, and the engine did not appear in any way to be thrashed.

The Manchester expresses are required to travel between Euston and Wilmslow, 177 miles, at a speed of 61.7 mph inclusive of speed orders at Rugby, Polesworth, Stafford and Crewe, to say nothing of a slight check at Norton Bridge on the up trip. The maximum load for one engine is 380 tons on these trains, but several times I have noticed that time has been recovered with loads in excess of this. Drivers Webster, Bowdon, Coombs and Rogers of Longsight shed, with loads varying from 404 to 420 tons, have been able to recover from three to five minutes on this tight schedule on various trips which I have made. Another fine performance was that of Driver Devereux of Bangor who, with 403 tons behind 6146, brought the 10.35am from Crewe to Euston, 158 miles, in 153 minutes.

A few weeks ago Driver Fairbrother of Liverpool, who had a load of 457 tons, made a valiant attempt to recover time lost by a spurt that would have accomplished the 46¾ miles from Bletchley to Euston, start to stop, in 49½ minutes, had not a signal stop outside the terminus spoilt the effort. One of my footplate friends is sure that the Royal Scots, with a load of 225 tons, would come from Rugby to Euston in 69 minutes, and thus rob The Cheltenham Flyer of its speed record. The Baby Scots also do well, but the new taper boiler variety of them have suffered from shortage of steam on several occasions, a defect now remedied I am told. The two-cylinder variety, despite their six-foot driving wheels, show a fine turn of speed on the Manchester expresses of the Midland section.

The Claughtons are disappearing, but in their day they did good work, though the Midland men never loved them much. One of this class, 5963, with Driver Cobb of Camden, reached Euston in 73 minutes from passing Rugby with no less than 390 tons. Another, 6016, with Driver Hawes of Edge Hill Liverpool, brought the 6.12pm from Crewe to Willesden in just 130 minutes with 230 tons, an average of just under 66 mph.

For the Midland compounds I have great respect. When they were on the Birmingham two-hour expresses I often noted their ability to take loads of 250-300 tons through Rugby in a minute or two under even time with drivers such as Preston, Hall, Copperwheat of Camden, and some of the Bushbury shed. No 1046, Driver Elliot of Derby, with a load of 327 tons, 57 above the maximum, succeeded in keeping time from St Pancras to Derby, 99 miles in 109 minutes.

I once had a footplate pass on one of these engines over this section, and was surprised at the easy way in which she steamed – though, as she was due for overhaul, her springing provided me with something of a liver tonic! I understand now why many drivers judge engines by their springing, and in this respect modern engines do not seem to have improved much, judging by the complaints I hear.

I have not the space to mention local journeys, but the new 2-6-4 type of tanks seem pretty popular and, unlike a similar class on the Southern now converted to tender engines, are said to ride well. Their work seems better than that of the older type they are superseding. The passenger does not get any experience of freight traffic, except when it blocks the line ahead! Judging from the number of old LNW "cauliflowers" one still sees, the British goods engine is a long-lived species. To the credit of the designer and the driver, be that said.

EDITOR'S FOOTNOTE

ESW sets out his easy relationship with express engine drivers, drawing on his logbook for past performances. The "cauliflower" was an 0-6-0 freight engine built in Crewe until 1902.

GWR's record of speed with safety

October 1935

The GWR people have a fine eye for publicity, and John Citizen accordingly has a notion that GW engines are the best. They are good, but so are others, and comparisons are impossible because of the very varied conditions of work. That the general design of that great engineer, the late G.J.Churchward, was thoroughly sound is shown by the fact that his successor enlarged Mr Churchward's types rather than altered the main design, and today is building Halls which, except for a smaller driving wheel, are of identical dimensions with the 2900 class, which first appeared over 30 years ago. No other company has been able to do anything like this.

Speaking generally, I have noticed better work by the Kings on the Plymouth than on the Birmingham route. If the train is not over 350 tons, the Stars, orn40s as the men call them, do every bit as good work as the bigger engines on the stiff Birmingham route.

The Bristol expresses are more often than not run by this class. The down trip includes a stop at Bath, 106¾ miles in 102 minutes, and my last three trips, two with Driver Cridlan and one with Driver Williams of Bristol, show a gain on schedule of four and two minutes respectively, with loads that varied from 244 to 398 to tons.

On the up train via Badminton there is no stop but a slack to slip at Reading. Here are three trips with drivers now retired from the link. Robson, net gain five minutes. Stephens, net gain four. Matthews, net gain four. Good work for engines built 25 years ago with loads of 285 to 388 tons.

The tight timing between stations on the GW calls for good powers of sprinting. Unluckily, despite the liberal allowances of station time, the station staff on the GW do not seem to hurry themselves to help drivers to save time, and so much of the value of the sprints is lost.

Here is a recent run with a King, 6016, from Reading to London, 36 miles start to stop in 35 minutes with 392 tons tare, and an engine that was finishing a 226-mile trip. The driver was Penfold of Laira, one of the oldest members of the Society. He retires this October, and the good wishes of all will accompany him on the conclusion of a very fine career. In distinction from many GWR men who favour a very early cut-off, perhaps 15 per cent, Mr Penfold favours 20 or even more and uses his regulator with discrimination.

That The Passenger is not solely interested in express work will be shown if I refer to a gallant effort to the credit of Driver Selwood of Bath Road shed Bristol who, with 211 tons and an ancient Bulldog with 5ft 8½in wheels, was faced with a schedule of 24 minutes for 18¼ miles from Bedminster to Weston-super-Mare, which includes a severe slack at the junction of the Weston loop. A signal stop outside Weston spoilt the effort, but had it not been for that our time would have been not more than 21 minutes, a sporting effort with such an engine.

Another sprint by Driver Wainwright of Old Oak Common shed, who had lost a bit of time by reason of poor coal, brought a 40 with a load of 300 tons over the stretch of 16¾ miles from Swindon to Chippenham in exactly level time; despite the fact that the grade is favourable, a 60 mph sprint over a short distance including start and stop is fine work.

The Castles are celebrated for their work with The Cheltenham Flyer, and I have logs of a couple of smart runs credited to drivers Bailey and Lewis, now both retired, which prove that the Castles still have a bit up their sleeve even on a 70 mph timing.

The GW are easily unrivalled amongst British railway companies in one important respect, that of safety, and though their extensive signalling devices are in part to be credited with this result, the men must have their share of praise. A while ago I went in a thick fog from Bristol to Bath, and the run was performed exactly on time. I happened to be in the front coach, and found myself praying that the electrical automatic signal was in good working order. As far as comfort in the coaches is concerned, the LMS and the LNER have the GW beaten, but when I think that the GW have lost only one passenger's life in the past 23 years I take my hat off to a unique record and those who make it possible.

EDITOR'S FOOTNOTE

In a first look at the GWR, ESW dips into his records for logs of trips before he became a columnist. His remark about time 'lost' at stations would not be his last on the subject.

Southern makes light of steep gradients

December 1935

The Southern does not catch the public eye with spectacular runs such as the other companies can stage, but the sympathetic passenger will admit that the work done by Southern locomen is not a whit behind that of their brethren on the larger lines. The King Arthur class – and especially the Glasgow-built engines, or Scotsmen – do work that only fine machines, well handled, could accomplish.

The Bournemouth expresses average just under 55 mph for the non-stop run with a load of around 330 tons, but there are severe gradients and a very bad slack at Southampton, as well as a necessarily slow approach from Clapham to Waterloo. The Bournemouth link who work these trains, Messrs Luke, Kerwill, Allen, Carter and Baker, to name those whose work I have observed lately, are good timekeepers, and I have yet to book lost time against the engine on any of these trips. The other night, on the 6.30pm from Waterloo to Southampton, we arrived ten minutes late, but as this included four pw slacks, signals and a dead stop outside Southampton on a trip that allows 88 minutes for 79¼ miles, 792 and Driver Baker may be congratulated.

The same great work is shown on the Exeter line, where the timings over some really ferocious gradients are amazing. Salisbury to Exeter, 88 miles in 93 minutes over a switchback course, including 2¾ miles of steady 1 in 80, represents as good locomotive work as any in the United Kingdom.

Before electrification the King Arthurs did some capital work on the Brighton line. With a load of 410 tons behind, 58 minutes for the 50½ miles is good going. My notebook credits that to 797 and Driver Baldwin, as fine a man as ever stepped on a footplate, now unhappily taken from us by death in the prime of his life.

The new Schools class are doing good work on the Portsmouth and Folkestone lines. The last run I timed on the latter was with 902, driven by Porritt of Ramsgate. A two-minute-late start from Folkestone was recovered by Tonbridge, where pw work cost us three or four minutes' delay. Though we were only two minutes behind schedule at Grove Park a suburban electric was put on in front, and we were 11 minutes late into London. To the passenger it seems an utter waste of coal and of the engine crew's labour to set a schedule that calls for almost a mile a minute from the coast to the outskirts of London, and then to nullify it by such a proceeding.

I am sufficiently aged to recollect the old days of the LBSC, even to trips behind the old signals, and, of course, Gladstones of Mr Stroudley's design. Was there ever a better designer? The famous "terrier" or "rooter" tanks were the Baby Austins of the loco world, whilst the D class of the 0-4-2 type have 60 years' record behind them, and some are still running. Just before electrification of the Eastbourne line one of them, called on owing to failure of a large engine, set out from Eastbourne with 249 tons and reached Haywards Heath, including the severe slacks at Lewes and Wivelsfield, in 36 minutes for the 28 miles. I had to check my watch by the station clock before I could believe the gallant little engine capable of such work.

Of the older types of SE&C engines I have some preference for the Germans, the 1770 class, though the other day one of Mr Wainwright's design, 1746, astonished me by reaching the stop at London Bridge in 13½ minutes after passing Orpington 12 miles out, and making the usual cautious approach all the way from New Cross.

I am living on the Southern, but our line is electrified, and I am too old to take on new interests. Electric motors leave me cold. I preferred the days of the Drummond tanks and the Jubilees. Some of the men from Windsor shed will have pleasant recollections of 111, a wonderfully free-steaming engine, and the Ascot shed of those days will not forget 126, the superheated member of the famous class, which gave some trouble with axle boxes but was a rare good puller, nonetheless.

Some of my old friends who once worked steam engines tell me they are getting used to electric traction and even like it, but I can't share their enthusiasm. I am with a driver of the old school who, asked if he would go on the electrics, said "No, I'd rather go back to goods; I am not a tram driver."

EDITOR'S FOOTNOTE

Individual engines, new and old favourites, surface here. ESW elaborates on Drummond tanks in his October 1956 column (page 125).

Locomotive Journal editors 1935-64

John Bromley, who commissioned ESW to write a column for the monthly *Locomotive Journal* in 1935 (see letter opposite) had been ASLEF general secretary and de facto editor of the *Journal* since 1914.

Bromley retired in 1936. From then until 1964, during which time ESW contributed regular articles, ASLEF general secretary/editors were:

1936 Richard Squance
1940 W.P.Allen
1948 Jim Baty
1956 Albert Hallworth
1960 William Evans
1964 Albert Griffiths

As the Editorial Notes from the May 1935 *Journal* reproduced here confirm, ASLEF was firmly on the left of the Labour Party. And, as ESW stated in his October 1937 column, he had no issues with that.

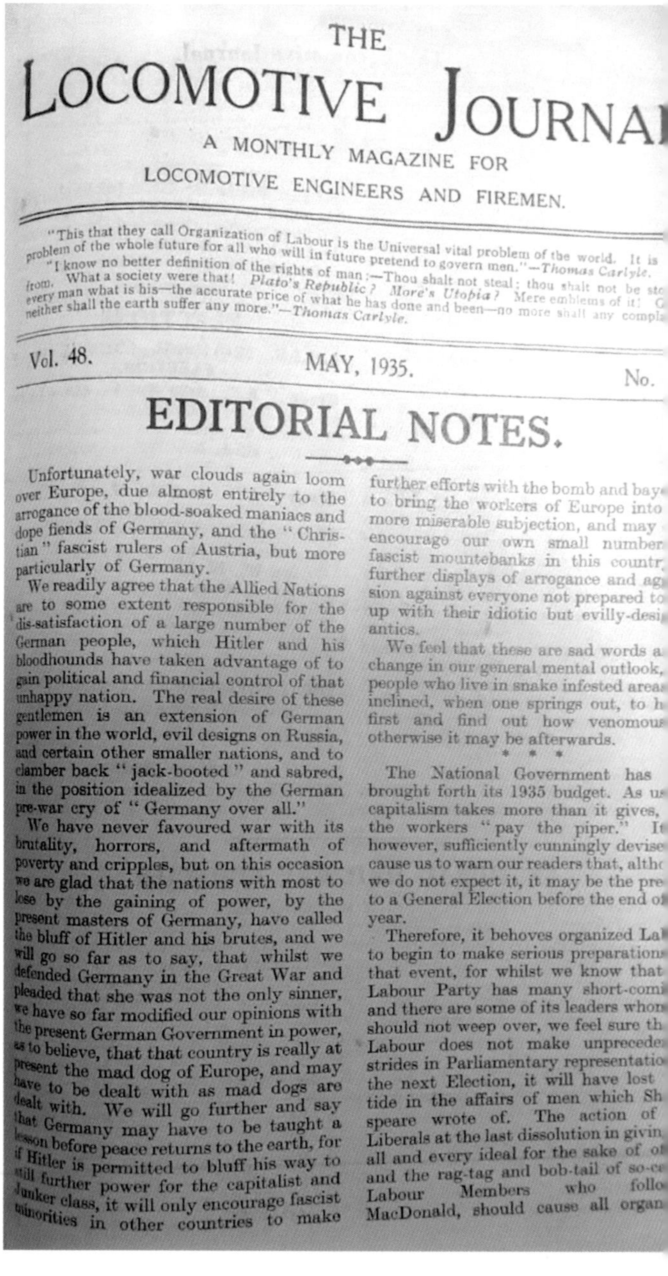

THE

LOCOMOTIVE JOURNA

A MONTHLY MAGAZINE FOR
LOCOMOTIVE ENGINEERS AND FIREMEN.

"This that they call Organization of Labour is the Universal vital problem of the world. It is problem of the whole future for all who will in future pretend to govern men."—*Thomas Carlyle.*

"I know no better definition of the rights of man:—Thou shalt not steal; thou shalt not be sto from. What a society were that! *Plato's Republic? More's Utopia?* Mere emblems of it! G every man what is his—the accurate price of what he has done and been—no more shall any compla neither shall the earth suffer any more."—*Thomas Carlyle.*

Vol. 48. MAY, 1935. No.

EDITORIAL NOTES.

Unfortunately, war clouds again loom over Europe, due almost entirely to the arrogance of the blood-soaked maniacs and dope fiends of Germany, and the "Christian" fascist rulers of Austria, but more particularly of Germany.

We readily agree that the Allied Nations are to some extent responsible for the dis-satisfaction of a large number of the German people, which Hitler and his bloodhounds have taken advantage of to gain political and financial control of that unhappy nation. The real desire of these gentlemen is an extension of German power in the world, evil designs on Russia, and certain other smaller nations, and to clamber back "jack-booted" and sabred, in the position idealized by the German pre-war cry of "Germany over all."

We have never favoured war with its brutality, horrors, and aftermath of poverty and cripples, but on this occasion we are glad that the nations with most to lose by the gaining of power, by the present masters of Germany, have called the bluff of Hitler and his brutes, and we will go so far as to say, that whilst we defended Germany in the Great War and pleaded that she was not the only sinner, we have so far modified our opinions with the present German Government in power, as to believe, that that country is really at present the mad dog of Europe, and may have to be dealt with as mad dogs are dealt with. We will go further and say that Germany may have to be taught a lesson before peace returns to the earth, for if Hitler is permitted to bluff his way to still further power for the capitalist and Junker class, it will only encourage fascist minorities in other countries to make

further efforts with the bomb and bay
to bring the workers of Europe into
more miserable subjection, and may
encourage our own small number
fascist mountebanks in this countr
further displays of arrogance and ag
sion against everyone not prepared to
up with their idiotic but evilly-desi
antics.

We feel that these are sad words a
change in our general mental outlook,
people who live in snake infested area
inclined, when one springs out, to h
first and find out how venomous
otherwise it may be afterwards.

* * *

The National Government has
brought forth its 1935 budget. As us
capitalism takes more than it gives,
the workers "pay the piper." It
however, sufficiently cunningly devise
cause us to warn our readers that, alth
we do not expect it, it may be the pre
to a General Election before the end o
year.

Therefore, it behoves organized La
to begin to make serious preparation
that event, for whilst we know that
Labour Party has many short-comi
and there are some of its leaders whom
should not weep over, we feel sure th
Labour does not make unprecede
strides in Parliamentary representatio
the next Election, it will have lost
tide in the affairs of men which Sh
speare wrote of. The action of
Liberals at the last dissolution in givin
all and every ideal for the sake of o
and the rag-tag and bob-tail of so-c
Labour Members who follo
MacDonald, should cause all organ

THE LOCOMOTIVE JOURNAL.

The Official Organ of the Associated Society
of Locomotive Engineers & Firemen.

P 11 / 141

EDITOR.
J. BROMLEY.

9, ARKWRIGHT ROAD,
Hampstead, N. W. 3.

Rev. Professor E.S. Waterhouse,
The College Villa,
RICHMOND,
Surrey.

10th December, 1935.

Dear Mr. Waterhouse,

I am writing to thank you very much for the series of interesting articles which you have so kindly contributed to our columns under the heading "As the Passenger sees it."

Your articles are causing quite a lively interest amongst our members, as it is something new to them to have the names of the enginemen given and with such kindly reference to their work. You may not know, but it is a very common expression amongst our members, that after a stiff trip, perhaps with fog or greasy rails, or other troubles, that the passengers generally pass out of the great terminals without looking to see whether it was an engine or a wheel-barrow which brought their train in.

I sincerely hope that you may be able to continue to write for us.

May I presume to wish you most hearty compliments of the coming Season, with best wishes for your health and general well being in the coming year.

Yours sincerely,

J Bromley.

EDITOR.

No margin for high-speed timings

February 1936

The average engineman realises that these are days of speed, and accepts the new tightly-cut schedules, if not with enthusiasm, with equanimity. Given an average load, fair weather conditions, a clear road and a good engine, no great difficulty is experienced in keeping time, even on the tightest of modern timings. What has not been sufficiently appreciated is the fact that there is virtually no margin for the various incidental hindrances that occur three trips out of four.

Now high speed may be a good advertising point but unpunctuality is certainly not, and if punctuality is to be expected a certain margin must be allowed. How seldom it is that a trip of 100 miles or so is performed without an adverse signal or a pw slack! If each time this happens late arrival follows, it is far better to allow a more liberal timing and keep strictly to it. The old days when a man had his own engine and knew all its moods and whims have gone, I fear for ever. But surely, even if the maximum mileage is demanded in the interests of economy, it is unfair to expect that any engine, often from a foreign shed, which may happen to be of the same type as that usually allotted to the train, will serve.

I reached Birmingham last month 12 minutes late, of which about six must be booked against the engine. The driver was a competent man from Bushbury shed, but the engine hailed from Longsight, had already come up to Camden, and was hustled to Wolverhampton and back before returning to Manchester. It was steaming poorly, and quite strange to both driver and fireman. The load was 293 tons, and the 112¾ miles, with a two minutes' stop at Coventry and service slacks at Rugby and Brandon, are scheduled to be done in 115 minutes.

On the return trip, another Baby Scot, again from a foreign shed, was worked to time by Driver Malkin of Bushbury. But it was evident to me, from the first coach, that the engine was worked hard all the way and steaming sullenly, so credit is due to driver and fireman for their work in keeping the schedule. Both trips would have presented little difficulty with an engine in good fettle, and one that the drivers knew.

A recent trip on The Bristolian, 118½ miles in 105 minutes with a severe slowing at Bath, confirmed my impression. The load was 217 tons, which is light enough for a King, though, of course, there is less braking power when travelling at high speeds, a fact that makes some men naturally a bit nervous. Driver Taylor of Old

Oak Common passed Didcot, 53¼ miles, in exactly 45 minutes, 46½ being allowed, but two pw checks made us nearly four minutes late in, as the schedule can scarcely afford time recovery to the extent of more than a minute or so.

The LNER Silver Jubilee flyer is in the hands of a link of eight King's Cross men, and confined to four specially-kept engines. The same applies to the Pullman expresses to Leeds, which are worked by a few men from Leeds and King's Cross sheds with Atlantic engines which are kept in good condition for this work specially. Consequently the time-keeping and recovery on both trains are good, despite the fact that the Atlantics are a 25-year-old engine, and the load 300 tons as a rule.

The British engineman can work the modern high-speed train safely, and to time, given a fair chance, but the modern notion of interchangeability of engines from sheds all over the line, whilst it may be all right for average timings, should not apply to the high-speed expresses.

I am sure that shed superintendents would agree with me, but I have a suspicion that their own views are overruled from headquarters by officials to whom an engine is an engine, just as a penny is a penny, who do not realise what kittle cattle steam locomotives are, and what enormous differences exist between two of a class, both as regards temporary conditions and more permanent differences that remain with an engine all its running days.

I like high speed as much as anybody, but it makes me positively uncomfortable to feel a harassed crew flogging a reluctant engine in a losing game against a severe schedule. I sometimes welcome a check which will give some excuse for the inevitably late arrival under such circumstances. We shall never get the best results from our high-speed trains until we give the driver fair play with an engine that is not simply theoretically, but actually, up to its job.

EDITOR'S FOOTNOTE

Speed, the main selling point of crack mid-1930s expresses, puts unfair demands on drivers unless engine and infrastructure are in top condition. It's a theme ESW will explore elsewhere. Baby Scot is the nickname of the LMS Patriot Class.

A chance to evaluate the Turbomotive

March 1936

Enginemen on all the lines in the country have read with interest of the LMS experiment in turbine propulsion. The steam turbine has been used for many years in marine engines, and many will recollect the queer-looking turbine engine that worked experimentally in 1926-1927 on the St Pancras-Manchester line. But LMS 6202 is of another order, the first non-condensing steam-turbine locomotive to work in this country. In other respects she is identical with the Pacific class 6200, in which series she is numbered.

Instead of cylinders she has two turbines, one for forward and a smaller one for reverse, neatly housed in special casings so that externally she looks much like an inside cylinder 4-6-2 type. She has a wide chimney to accommodate the double blast pipe made necessary by the softness of the exhaust. The regulator is of a normal type, but instead of the usual process of notching up on the reverse the turbine engine has six nozzles, all of which can be used at starting, and when the engine has got under weigh some nozzles can be cut off, the number varying, of course, according to conditions of running.

The reverse turbine is separate and smaller, and has three nozzles only. Most of the difficulties that have been experienced with the engine are due to the reversing turbine. Indeed, on the occasion of the trip I made I left the engine at Euston still in forward gear, all efforts to get her into reverse having failed. But this is an incidental defect and can no doubt be overcome.

The train on which I travelled was the 9.08am from Crewe. The engine was excellently handled by Driver Lyon of Edge Hill, with Driver Taylor of the same shed on the footplate as instructor. The load was 390 tons tare. The first stop was Bletchley, 111¾ miles, 118 minutes being allowed, with slacks at Norton Bridge, which is but a slight one, Stafford, Polesworth and Rugby.

Leaving Crewe 3½ minutes late, we glided away as smoothly as an eight-cylinder car. I was in the first compartment with Driver Johnson of Crewe, who was travelling as a passenger, and we were both struck by the silence of the exhaust and the smooth pull of the turbine as compared with the ordinary sound of the steam at starting and the slight jerking of a reciprocating piston. The day was foggy, and at one or two points such as the junction at Colwich and outside Rugby the engine was eased, presumably because of the difficulty in seeing signals. Rugby, 75½ miles, was passed in 75 minutes, ahead of time. The Mancunian express, which does not stop at Crewe, is allowed 74, and the fastest timing, that of the 6.12pm from Crewe, is 73, but the load is lighter.

A dead stop for signals at Castlethorpe and further signals at Wolverton brought us to Bletchley two minutes late. We left three late, but the 15 miles to Tring summit were done in 17 minutes and the remaining 31½ in 31. Indeed, the engine was worked quite easily after Tring, and we arrived nearly a minute ahead of time.

It was a most interesting trip and showed that the Turbomotive is fully capable of handling any normal load on a mile-a-minute schedule. How far she will prove an economic success I do not know. She cost £14,000 compared with £9,000 for the ordinary members of her class, though she should save on coal and, when she is stabilised, on repairs. Moreover, she is light on the permanent way, owing to the absence of hammering action of an ordinary engine. She carries an axle load of 24 tons on the latter pairs of coupled wheels with a total adhesion weight of 70 tons 13 cwt out of her 100 tons total. That should be an advantage both in regard to traction and to the road itself.

As the passenger sees it, outwardly this is a neat-looking, characteristically British engine. From the train one notices the quiet and smoothness of the motion, which I understand is felt also on the footplate. That will be welcomed by drivers and firemen who sympathise with one of their number who told me he was thinking of applying for a job as a cowboy since he was sure he could ride any bucking bronco after experience with his engine. So here's success to 6202, and if they give her a name may it be easier to handle than some of the latest, for example Saskatchewan and Tanganyika.

EDITOR'S FOOTNOTE

6202, the Turbomotive, proved to be the only one of its type built. The Second World War put a stop to experimental design. In 1952, 6202 was rebuilt as a conventional Princess Royal class Pacific and named Princess Anne, but after only two months of service was among the engines involved in the Harrow & Wealdstone disaster, and scrapped as unrepairable.

Journal letter, April 1936: The Passenger's perceptive plea for regular engines

SIR – It was a pleasure to read, in the February 1936 issue, our contributor Mr E.S.Waterhouse who, although writing "As the Passenger sees it", made a plea for men to have a more regular engine that would be hard to beat as a presentation of the case.

I have waited for the LMS annual meeting knowing that Sir Josiah Stamp likes to present some locomotive statistics. After stating that the 10,316 engines were now reduced to 7,885, and types from 393 to 185, he said that when present orders are completed types will fall to 150 and the stock to 7,554.

Last year train miles increased to 2,750,000, double-heading was reduced by 20 per cent, shunting mileage attained its lowest figure, coal consumption per engine mile continued to fall, with a reduction of 19 per cent in engine casualties, the number being the lowest recorded by the LMS.

From 1931 to September last 793 new locomotives were placed in service on the LMS, with its route mileage of about 7,000, and this exceeded the number of new locomotives, steam and electric, put on the United States railways, with a route mileage of 250,000, in the same period.

Among the items claimed as responsible for this is the better scheduling of engine workings. It looks very nice on paper, and presents a picture of economy with efficiency that should gladden the heart of any shareholder, but the point is that an even greater economy, and certainly efficiency, could be attained if the nearest approximation to regular engines could be worked out with the schedules.

It is not easy or possible in all cases, but given the desire by those "high officials to whom an engine is an engine just as a penny is a penny" (vide Mr Waterhouse) the footplatemen, to whom it would mean so much, will not be backward in coming forward with helpful suggestions.

I should be interested to know of any large depot where men have their regular engines today in any group, apart from those mentioned by our observant passenger friend.

J.BIRCH

EDITOR'S FOOTNOTE

As LMS chairman, Josiah Stamp appointed W.A.Stanier chief mechanical engineer in January 1932.

My method of comparing journeys

April 1936

To me a train journey is always interesting because, in all journeys by express trains, I keep a log of the trip. First, class and number of the engine, then the name and shed of the driver, then the load, and then a note of weather conditions. I enter the times at various points on the road, noting signal and pwr checks and, on getting home, record it in the book I keep which gives my other journeys over the same route.

Looking over that route, three things stand out. First, the regularity of running times, so much that I can usually forecast to a minute what time we shall pass the next timing place, although now and then an unexpected spurt or falling off puts my calculations out.

Second, the rarity of having to book time against an engine; if that happens with anything up to the maximum load allowed, poor coal is almost always the reason, though the present "economy" in lubricating oil, together with the system of interchange of engines all over the place, leads occasionally to some trouble with valve gears.

The third thing is that on average one check, signal, pwr or something similar, takes place about every 60 miles. It is hardly ever that any long run is performed without several such delays, and this is easily the chief source of unpunctuality.

A fob watch used by ESW to time journeys. He typically carried two such watches, one in each waistcoat pocket, to verify timings

EDITOR'S FOOTNOTE

The book containing ESW's logs over the years has not survived. For some passengers, train timing was a hobby; to him it was the nuts and bolts of his craft.

The station pilotman's story

May 1936

The station pilot, or banker, stood at its accustomed leisure in the bay, and I, unemployed as I was until my connection was due, greeted the driver with the remark that his was a dull job.

"Well", he reflected, "there's some that don't like a waiting job, and some that don't mind. I'm one of the sort that takes things as they come. Anyhow, I'd rather have this than a double-home job."

That led to some talk about double-home trips and the respective merits of company barracks and lodgings.

"Well", said the pilotman meditatively, "there's two sides to that question as the wife of one of my mates knows." Scenting a story, I asked for further information, and this is what I heard.

"Tom Steamchest and his wife were a happy couple. Steady chap, Tom, sort of man who never forgot the little things that please women. Bit of a model husband. Mrs Steamchest proud of him, too.

"Well, everything went along happily until Tom got on the London link, and that meant a double-home job. Mrs Steamchest got a bit lonely. They had no family and she found things pretty quiet, you see. Now Tom, as a home bird, didn't like barracks, and got lodgings with a widow woman. Cost him a bit more, but Tom liked to be comfortable.

"Things went on all right till one day Mrs Blower, wife of Dick Blower, Tom's mate, got talking. She asked Mrs Steamchest why Tom couldn't sleep at the barracks, same as her man. Of course it wasn't her business, but she'd see her husband didn't go staying at another woman's house on his own, and all that sort of thing.

"Well, I suppose it got on Mrs Steamchest's nerves or something. Anyhow, she began to worry about it and, at last, without saying a word to anyone, she determined to go up to town and spy on poor old Tom. So off she goes, and one evening found herself in the street where Tom lodges, looking round the corner at the house, and wondering how she could find out anything about her man.

"Suddenly, to her surprise, she sees Tom come out, neatly togged up, and walk down the street. Mrs Steamchest follows him. Presently Tom goes to a telephone box, and Mrs Steamchest thought hard, I can tell you. Tom comes out, walks along, and stops at a jeweller's, and after looking in the window a bit, goes in.

"Of course, that proved things to Mrs Steamchest. She followed Tom, saw him go to a house, knock and enter. It was dark, and Mrs Steamchest slipped into the front garden and tried to look through the blinds. She saw nothing, and presently Tom came out again and set off. Mrs Steamchest hid behind a bush till he had gone and was about to follow him when a hand fell on her shoulder and a gruff voice asked her what her game was.

"To her horror, Mrs Steamchest found herself facing a policeman. Foolishly she got panicky, wriggled from his grasp, and tried to run away. She was caught in a moment and hauled off to the police station. What would have happened I cannot say, but luckily she thought of Dick Blower at the barracks. They sent for him. Dick came along and, to him and a greatly amused sergeant, Mrs Steamchest sobbed her story. Of course they let her go. Dick took her to the station and put her on the night mail for home, promising not to say a word, especially not to Mrs Blower.

"Next day Tom came back as usual, kissed Mrs Steamchest heartily and said 'Have you forgotten our wedding day? I didn't. Last night I had rotten toothache. I went out and phoned the dentist, and he told me to trot along to his private house as he was just leaving the surgery. On the way I saw this in the window and brought it along for you, old girl.' Tom produced a nice little gold broach. 'Then', he said, I went along to the dentist's and had a tooth out. Anyway, that's one beggar less to ache.'

"Poor Mrs Steamchest began to cry. 'Hold hard, old girl' said Tom. 'Is that what you do when I bring you a present?' Mrs Steamchest stammered something about being sorry for poor old Tom's toothache, and I suppose old Tom thought what a nice, sympathetic wife he'd got. Anyhow, Mrs Steamchest left off worrying about Tom and one day he took her up to town and introduced her to his landlady, who was a good old soul and a grandmother."

"How did you know all that?" I asked.

"Well, you see," said the banker's driver, "my name happens to be Blower. Tom's finished now, and he and his missus have gone to join her sister in Canada, so I suppose there's no harm in letting on, and anyhow you don't know Tom."

"That's right", I said, "and do you still sleep in the barracks?"

"I told you", he replied, "there are worse jobs than the station banker."

EDITOR'S FOOTNOTE

A cautionary tale of double-homing, an issue much-debated in ASLEF circles during the 1930s. Was it wise to call the suspicious wife Mrs Steamchest? The station pilot was maintained at principal stations for relief work and shunting.

The Silver Jubilee and other expresses

June 1936

My first trips on The Silver Jubilee took place recently. As is well known this, Britain's first streamlined train, runs from King's Cross to Newcastle and back in four hours each way, including a two-minute stop at Darlington. The distance is 268¼ miles and the average speed works out at a fraction over 67 mph. The load is 220 tons tare.

Driver H Gutteridge with Silver Link 2509 was in charge of the up journey. The first impression was the amazingly quick start made from cold. Birtley, nearly 5½ miles, was passed in 7½ minutes and Darlington reached a minute early. Leaving there punctually, the 44 miles to York, where there is a severe slowing, took 39 minutes.

The first pw check, excepting those more or less continual over some colliery workings in the earlier part of the journey, was at Bawtry. Driver Gutteridge had prudently got a minute or two ahead, recovered smartly, and was still ahead of time at Grantham. No exceptionally high speed was made down the bank from Stoke summit, but we continued to keep a couple of minutes in hand for another pw at Biggleswade.

Hitchin was passed on time and then followed an amazing rush up the bank to Woolmer Green and the highest speed, just 92 mph, on the falling grade after, so that the distance of almost 14¼ miles to Hatfield was done in 10¼ minutes. This was to get ahead for the severe check imposed by reason of a bridge being rebuilt at Barnet, and we rolled slowly into King's Cross to time.

It was a remarkable feat of good driving, each check being anticipated, and the impression one got of the easy mastery of Silver Link over her load. The engine never seemed to be pressed. There were no astounding speeds down hill, but extraordinary liveliness in acceleration and climbing. Driver Gutteridge retires in July, and when he goes the company will lose a master man, worthy of the best traditions of the footplate.

On the down journey I travelled by the afternoon Scotch express which was loaded to 510 tons. Driver Hutchinson of Gateshead was in charge, the engine being Manna (2596). The bridge slack lost us five minutes and we were three late into, and six late out of, Grantham. Four minutes were recovered to York and the excellent way in which the engine was worked made it possible for the remainder to be gained between Darlington and Newcastle, a further example of the wonderful hauling capacity of the LNER Pacifics.

Of their sprinting powers I had evidence on the Saturday before Easter when I travelled by the 8.40am from Grantham, which gets 100 minutes for the 105¼ miles. The engine was Gainsborough (2597}, temporarily shedded at Doncaster, with Driver Shooter, one of the younger men not yet in the Pacific link. As it was a relief train the load was light, 217 tons, which enabled us to climb to Stoke summit in 7½ minutes. From there followed an amazingly fast run, the 24 miles to Peterborough, passed at 20 mph, being done in 17½ minutes as compared with 19 minutes on the Silver Jubilee run. This enabled us to pass Hatfield four minutes ahead of time, and despite the bridge check at Barnet and signals from Wood Green on we reached King's Cross just to time – a smart performance which shows that there are men coming along to maintain the traditions of the Pacific link at Doncaster, which I reckon as one of the best in the country.

A trip on The Mancunian recently enabled me to travel behind the new Royal Scot, British Legion (6170). As the load was 46 tons over the limit of that train – 380 tons – we had a taper boiler Baby Scot as pilot, Driver Rowbotham of Longsight being on the train engine and Driver Tharme of Camden on the "hooker". The two engines made nothing of the load and despite two checks we were in four minutes early on the 61 mph schedule from Wilmslow. I reckon we'd have kept time without that hooker.

I heard the other day of a driver newly appointed to a top link who found his engine decorated with a large L when he took it over. We evidently have some motorists in our sheds! So we have amongst journalists. LNER men will be interested to read in a recent newspaper article that the reporter was writing as the Flying Scotsman, having streaked like a comet through Peterborough, was speeding towards Grantham at 80 miles an hour. I should like to know what happened to the driver who does the comet business through Peterborough. I imagine that he'd now be playing a harp instead of pushing a Pacific, and much as I admire these engines I have yet to time them at 80 miles an hour up to Stoke summit with the Scotsman's load. But trifles like this are nought to our brave journalists!

EDITOR'S FOOTNOTE

Contemporary press exaggerations about speed are ridiculed by ESW, quoting route realities LNER enginemen would know well.

Three accounts of locomotive practice in North America

At a running shed in Colombus, Ohio

July 1936

A recent trip to the United States and Canada gave me a chance of seeing a little of footplate work and conditions on the other side of the Atlantic. On a broiling hot afternoon in May I visited the running shed of St Clair, Columbus, Ohio, where I was most kindly received by Mr Joseph J Sosey, the assistant roundhouse foreman. This is practically a 100 per cent Union shed.

In the States and Canada the Union is called the Brotherhood of Locomotive Engineers and Firemen, and surely there is much to be said for that name. In the States freight traffic counts for much more than passenger, and the majority of engines stabled at St Clair were freight types.

Rates of pay are complicated, depending not simply on the standard wage, overtime and mileage, but also on the class and weight of the engines. For example, men working the huge Mallett engines draw correspondingly higher pay. The standard rate for drivers, engineers as they are called out there, varies from $6 72 cents to $7 70 cents a day. A dollar is of course 100 cents, and $5 go to the English pound. Firemen get from $4.90 to $6.19. Overtime is paid either on a time basis of eight hours or on a mileage basis of 100 miles and at the rate of time and a half. Retiring age is fixed at 70, and a pension is then given. Freight men receive rather more than passenger men, but they are called out in turn as their jobs come in, whilst the passenger men get regular rostered turns, so that on the whole the advantage is with them.

Even if we allow that living, particularly rent, is dearer than in England, there is no doubt that our American brothers are more adequately paid than their English counterparts. I saw a car park at the sheds where a number of cars awaited locomotive men who had driven up to the shed and gone on duty, and though cars are cheaper to buy and run in the States than here, still we are a long way from car parks at British engine sheds.

Discipline is strict. Men on duty, of course, must abstain from intoxicants. Many American lines, however, expect similar abstinence off duty, even now that prohibition is repealed. It will be realised that there are still individual "dry" states and no uniform liquor laws for the whole country.

The safety record of American lines is magnificent. Last year not a single passenger's life was lost in a railway accident. When we remember that there are something like a quarter of a million miles of railway and many scores of separate companies, this record is marvellous. On some lines it is greatly aided by mechanical devices. On the New York Central, for example, the main lines are fitted with electrical safety plant. In fog the engineman can sit comfortably in his bucket seat and watch a little panel on which electric lights indicate the position of the signals.

The work of the fireman on the bigger engines is made light by the mechanical stoker. A worm-screw works the coal up from the tender, and powerful steam jets, operated by handles in the cab, blow the coal to different parts of the firebox as desired. The coal is, of course, pulverised before being placed on the tender. Some of the tenders carry 14,000 gallons of water and 25 tons of coal. Some tender!

A typical American locomotive man wears a soft blue and white striped peak cap, overalls with shoulder straps, gloves and glasses. The dusty coal makes the last a necessity, and they protect the eyes from glare and wind.

One trouble is the very varying quality of the water. In some sheds the boiler is emptied after each trip, and even then a considerable scale keeps forming. There are pick-up troughs on some lines, but they are less in evidence than in this country, and in any case there are less long-distance non-stop runs. That makes it possible for most men to return home the same day, though the double-home system is not unknown in the States.

The engines frequently do far longer trips than the drivers, and this makes it impossible for the same men to have the same engine regularly, though I was told that the advantages of this system are not overlooked, and where possible it is employed.

I left the Columbus sheds having cooled down by means of a draught of ice water, supplied from a tap for employees. Think of the lukewarm stuff that comes from an English tap on a hot day!

EDITOR'S FOOTNOTE

ESW was in North America as a conference delegate. These observations focus on the working conditions of the US engineer and fireman. They spell out how well-organised American union brothers were in 1936, and emphasise the overall safety record of railway companies.

***Journal* letter, August 1936: American innovation benefits**

SIR- I was delighted to see the writing of a Mr E.S.Waterhouse in the issue of July 1936 for I thought surely we have at last found a man to take the place of the late Clement E.Stretton. Some of the boys like myself will remember him; he resided at Leicester, and was consulting engineer for the Society,and was also a lawyer and did good work for the Society. He wrote with an open mind and tried to infuse knowledge to those who read, I think. A thing much wanted in the coming age amongst every class of workers.

I have for 50 years read the monthly journal of the American Brotherhood of Locomotive Engineers, and also the Brotherhood of Locomotive Firemen's and Enginemen's magazine and have been surprised how certain fittings have been put on British locomotives after a visit to the USA by officials of our railways, 20 years after the American roads had them.

Don't forget that they were the first to use the Westinghouse brake, etc. Influence by the locomotive brotherhoods has caused their companies to be the first to put cabs on engines. Pardon me for not saying anything about things going from this country and being adopted there as an improvement. I am glad to say there is admiration for one another in Britain and the USA. and such writers as Mr Waterhouse will help things.

E.S.Spencer

To Montreal on the International Limited

August 1936

From the States I passed to Canada at Niagara and made my way to Toronto, joining the International Limited there, although it runs right through from Chicago. Leaving Toronto at 9.15 am, it is due in Montreal, 312½ miles away, at 5.15 pm, which does not seem a fast schedule – but as it included 20 intermediate stops and one service stop to take up water, and we did not leave until 9.52 and arrived to the minute, it will still be seen that fine loco work was involved.

We started behind a huge electric locomotive, changed after a few miles for Canadian Pacific 4-6-4 K5 type, No 5702, which went right through, though we changed crews twice on the trip. The engine was mechanically fired and fitted with a booster. The load varied a lot en route, but was for the most part 17 cars, the lightest being about 50 and the heaviest 80 tons. Exact load is not given in the waybill, but it must have been between 650-750 tons most of the way.

The road is flat and pretty straight, running by the side of the lake for the greater part of the way. The tractive effort of the locomotive, printed on the cab side, was 43,000 tons compared with 40,000 tons of the LMS Pacifics and the GWR Kings. We cannot flatter ourselves that either of them could have shown to much advantage with such a load and so many stops. One difference lies in the big boiler which makes the nominal tractive effort represent so much more haulage capacity. A further advantage is the booster, which enables a quick getaway to be made after each stop. One stretch of 30 miles was covered in 26¾ minutes start to stop, averaging just over 67 mph, and yet the maximum speed I timed was 78 mph. The booster was clearly responsible for the overall average attained.

At Montreal I had a day at the Angus works of the Canadian Pacific by courtesy of the locomotive superintendent, who recruited Mr R A Smith of the mechanical department to show me round. Space does not allow a description of the up-to-date plant of this huge works or of the varied types of engines being built and repaired. One interesting fact is that firemen have a period of work with the fitters, which gives them an opportunity of getting real inside knowledge of the construction of the engines.

The visit of the Royal Scot seems to have left one result at any rate, as the latest express engines of the CPR are being painted in LMS red, an innovation in a land where dull black is the rule for engine painting.

What was particularly interesting were the blueprint drawings supplied to me of a new semi-streamlined 4-4-4 express engine, a small machine according to Canadian ideas, built for speed. The chief dimensions, possibly to be modified before the actual engine is turned out, are as follows: driving wheels 6ft 8in; cylinders 17¼ x 28; boiler pressure 300lb; combined heating surface 3,933 sq ft; grate area 55.6 sq ft; loaded weight of engine 265,000lb; of tender 201,000lb; capacity of tender 7,000 gallons of water and 12 tons of coal; weight on driving wheels 120,000lb; tractive effort 26,500lb. It will be interesting to see the results of this experiment, which suggests that CPR have in mind some speed trains of light load.

EDITOR'S FOOTNOTE

How huge North American locomotives generate the power to service long-distance routes with many stops, plus a pointer to a smaller, faster version.

Transatlantic comparisons are worthless

September 1936

Any comparison between British and American locomotives as to which is best is silly, for under English conditions the American locomotive would be extravagant and inefficient, while the English locomotive would be useless in America. We build small engines of great efficiency to carry limited loads at high speeds. In America loads are enormous, and there are far less long-distance non-stop runs. What is needed is power and rapid acceleration, and this is got, though at the price of a big fuel consumption.

The tank engine is relatively as rare in America as it is common here. Shunting in the yards is done by 0-6-0, sometimes 0-8-0 or even 0-10-0 tender engines. The freight traffic of the railways is more important by far than the passenger traffic. Freight trains are composed of huge covered cars, far larger than ours, and all are Westinghouse braked.

Speaking of braking reminds one that there is far more jerking at starting than in the case of English trains. It is not altogether fair to blame the American engineer for this, though some of those who drove me were far more skilful at starting than were others. In part it is due to the fact that trains are more loosely coupled than here, and also to the fact that with long trains it is not always possible to get the brake fully off the end coaches before the start is made.

The steam locomotive is disappearing from the great towns. For example, all passenger trains leaving New York termini are electrically hauled. In the case of the Pennsylvania line on which I travelled, the electric locomotive went as far as Paoli, 112 miles from New York.

In appearance we have the Americans well beaten. Their locomotives, painted dull black or olive green, look very undistinguished. They are not lined out, and save for the name of the company to which they belong, there is little to distinguish those of one line from another. Moreover, the designs are less neat and the engines look out of proportion.

Yet in many respects the Americans are in advance of us. The oil lamps used on British railways might surely be replaced by electric lighting. In the Angus works at Montreal I saw a neat little device by which an electric light could be focussed on any gauge in the cab, giving instant vision, and then as easily be switched off.

Certain features seem uniform throughout America and Canada. Practically all locomotives are of the two-cylinder type with outside cylinders, at least one may say that is the rule. Another typically American feature is that of pitching the boiler high over the wheels, with daylight showing between them. The most British looking engine I saw belonged to the Delaware and Hudson line. It had a much neater external appearance and smoke shields similar to those on the SR and LMS. This engine was fitted with roller bearings, and the engineer told me they were a great success.

There is an impressive sense of power and massiveness in the American engine. Our own Royal Scot, which was well-remembered since Driver Gilbertson so ably demonstrated its powers around the States, was greatly admired for its neatness and efficiency – a "real cute 'lil gadget" was the description given to me. But some American engineers are naturally annoyed at the silly talk of the Press in this country which suggested that the King George V and Royal Scot frightened them by their speed. Our shunting friends of Fleet Street might recollect that the record of The Cheltenham Flyer as the fastest steam train in the world has now been captured by an American train.

What we can claim, however, is a far greater number of long-distance non-stop runs than America boasts. Really express trains in America are few and far between, simply because there is a much less densely packed population in that great continent, and it does not pay to run intermediate services between towns to feed services calling only at the large places.

EDITOR'S FOOTNOTE

No direct comparisons, but differences between locomotion in North America and Britain abound. ESW usefully underlines the arrival of electrical power in the big conurbations.

Home again via Greenock and Glasgow

October 1936

After my experiences across the Atlantic it felt good to see a British engine again, actually No 662, an LMS standard 4-4-0, which took the boat train from Greenock, and I got so interested in chatting with the enginemen working her that I had to sprint hard to get back to my compartment in dire peril of being left behind.

I came from Glasgow to St Pancras. A Royal Scot, 6161, driven by Rae of Carlisle with 244 tons, had difficulty in keeping time – only the difficulty was in not getting in too soon for we were steaming easily and getting ahead of schedule all the way.

A taper boiler 4-6-0, 5618, dropped a bit of time up the Aisgill summit, the load being now 284 tons, but recovered it by hard running from Hellifield to Leeds. From thence a similar engine, 5664, worked by Crow of Leicester, took us along with no loss of time despite four pw checks.

Driver Irish of Kentish Town took charge from Leicester and we made London to time despite a three-minutes-late start and signals at Hendon, smart work on a schedule that calls for 90 miles in 108 minutes.

I have not done a great deal of travelling since my return, but as one who is interested in good loco work, not simply on fast express turns, I should like to hand on a couple of diplomas. The first is to Driver Root of Bricklayers Arms for punctual arrival at Charing Cross on an up Hastings train after several signal checks. In this case the engine was 939 of the SR Schools class. The second goes to Driver Macdonald of Bedford with compound 1070 and the heavy load of 340 tons on a local train from Bedford to St Pancras. Such a load wants starting with an engine of that capacity.

The best bit of work I recorded in August was done on the 2.30pm from King's Cross to York one Saturday. A couple of Atlantics took the train, which had a tare load of just under 500 tons, to Grantham easily enough. The 82¾ miles from there to York are to be done in 92 minutes. One of the new batch of K3 mixed traffic engines, 2451, which I think belongs to Peterborough, took charge of the train and was handled by Driver J.Clough of York. I was not prepared for a 2-6-0 engine with such small wheels to prove capable of handling such a load at such a speed.

Starting a minute late we got away so well that Newark, 14½ miles from Grantham, was passed in just over 16 minutes and Retford, nearly 19 miles further, in another 17 minutes. Outside Doncaster two sets of distant signals were on, which cost us between three and four minutes, but a gallant recovery brought us into York a minute ahead of schedule.

Steam pressure was at the blowing off point nearly all the way. No doubt the big boiler with which the K3 class is furnished was responsible for the adequacy of power, but the speed was a surprise which made me wonder if my idea that 6ft 6in is a minimum for the driving wheel of an engine used for fast work is correct.

Returning via the Scarborough Flyer I found my old friend Stephen Maw of Doncaster and the engine of that name, a Pacific, at the head of the train. The load was light, 232 tons, and though the schedule calls for the 188 miles to be non-stop in three hours no difficulty was found in timekeeping despite the recovery of a three-minutes late start.

Mr Maw will have retired before these notes are in print. An old and well-known member of the Society, I have often travelled in trains driven by him. He is not one who delights in spectacular feats but a careful, steady driver who keeps to his schedule, running generally with a minute or two in hand, and I at any rate have never booked any time against him. The Doncaster Pacific link is composed of as good a group of drivers as any in the land, and Mr Maw has for several years been a distinguished member of it. He tells me that he has had enough and is looking forward to a rest. May he enjoy it long and thoroughly.

EDITOR'S FOOTNOTE

This account of the Glasgow-St Pancras route via the Carlisle-Settle line illustrates express alternatives open to LMS passengers in the 1930s. The column ends with one of several ESW tributes to veteran drivers on retirement.

The eyesight test and other grouses

November 1936

Like most of us, enginemen can "grouse" on occasion, and as an outsider I ought to be impartial in my judgement of the genuineness of the grouses I hear – that is, if it be true that the onlooker sees more of the game than the players.

Sometimes I am inclined to whistle the grumblers offside. For example, it is no use grousing about faster schedules. Railway transport is fighting for its life against motor competition, and faster passenger and freight traffic is the only way of competing with road vehicles. Most drivers have no objection to higher speeds as such. What they rightly feel a grievance about is the rough engines, overworked and under repaired, with which they are often expected to perform feats of which only an engine in good trim is capable.

One of the link who work The Silver Jubilee told me that he would rather take that train than an ordinary express simply because the engines are kept in good trim, the load is well within their power, and two sections ahead are kept clear, so that there is not the ever-present fear of passing the "sticks". But, as one of my friends remarked the other day: "When you don't know what your job may be when you report for duty, when you don't know your guard, nor your engine, your work may be exciting but it isn't pleasant."

Another grouse with which I have much sympathy is that of tools. A fireman leaves his engine in the shed properly equipped with lamps and tools. When he goes to book on again, he finds another engine entered for him, for, of course, in the meanwhile his former engine has been sent off anywhere, and on examining it finds it minus lamps and at least half the tools. He asks where he can get what he needs, and is told "Go round and pinch them!"

In the old days a man, having his own engine, could leave all his belongings. Some lines used to provide lockers where necessities could be stored. To have to scrounge round other engines to get tools is a poor business and, since no one is responsible, is certain to be a wasteful affair. It would repay the company if they arranged for someone to be in charge of such equipment and see that it was kept in good order and properly distributed.

This might provide a job for the man whose grievance I feel is a serious one. I refer to the driver who is unlucky enough to fail the eyesight test. I mention no names, but some who read this will know to whom I refer, and there are many like him. Driver X has done fine service on the mainline expresses. He is now sent to work as a shed labourer at a depot miles away from his home and at a wage not more than half what he was averaging, and is apparently to think himself lucky too.

Eyesight is the trouble. Now, the man whose sight is really bad cannot continue his driving. Yet I am not convinced that tests such as those given with a lantern are a true criterion.

Surely the point is: can he see signals day and night, not whether he can distinguish colours on the Edridge-Green apparatus. If spectacles are no disqualification for holding a motor driving licence, must it be held that no engineman who needs them is fit to be on the footplate? The answer given is that they might break or get clouded with steam. But a second pair could be carried to meet the first objection. As regards the other, every engineman I saw in America wore goggles, as do many over here. What about steam obscuring them?

I don't mean to make wholesale charges against officials, many of whom try to show considerateness in such cases, but I did hear one say that since there were so many younger men waiting for promotion, it was well to be strict with the eye test. That seemed to me quite unfair. It is a poor thing to do justice to the younger men by doing injustice to the elder.

The last grouse is my own. Unfortunately it won't reach the eyes of those who should hear it for it is this – "Why don't many men read their *Journal*?" These articles have been appearing for nearly two years. Yet I find men who have never noticed them. I don't ask anyone to read my stuff specially, but the man who has never seen it cannot have opened his *Journal* much. What's the good of expecting the Society to represent your interests if you don't read what they are doing and what they want you to do? Put the *Daily Stunter* in the firebox, and get the *Journal*.

EDITOR'S FOOTNOTE

A trenchant column. Unfortunately, no record exists of reader response. Highlighting the vulnerability of even top drivers to demotion for failing a controversial eyesight test, ESW demonstrates personal convictions about social justice.

My talk with a trades union doubter

October 1937

I have been on holiday and away from the train for a month. True, I wandered down to the quiet little station at Midhurst, saw a Southern D tank or two, those gallant veterans of the old LBSC, and had a chat with the Fratton crew who had just worked in a one-coach train with the old SW Jubilee No 627. Otherwise I have not smelt the pleasant odour of smoke and steam for the whole time.

This gives me a chance to report an interesting chat with a very decent fellow, a keen engineman, who is a non-unionist.

The first reason he gave was that he could do better for himself outside a union. This, I tried to show, was false. Many a man has been acquitted on a discipline charge, or has got compensation for an injury, because of the skilled assistance at his disposal. Moreover, even if it were true, it is a selfish argument.

No one can argue that the position of enginemen would be what it is today if they were not, as a body, well organised. I pointed out that he shared in the benefits that organisation had brought, but did not contribute to them. Then again, even if a man is right in saying (though I dispute it) that he can do what is necessary for himself, surely we ought to stand by one another and not judge simply by a narrowly self-interested point of view.

A second reason was political. My friend has been a lifelong Liberal and refuses to have anything to do with the Labour Party. I told him that I was brought up as a Liberal, and for many years was a keen supporter of that party. Now, owing to the changed conditions to which the wartime alliance with the Tories largely contributed, the Liberal Party had ceased to be a considerable force in political life. But surely what was best in that party belongs now to Labour. I cannot understand a true Liberal turning Tory because his own party has ceased to be represented in the constituency in which he votes.

The Tory Party is fond of paying lip service to the people and to democratic principles, but it represents a class of the community whose interests are in capitalism, big business and landed proprietors; and the workers, whether they work at clerical or mechanical tasks, are acting against their own interests in supporting Toryism. I am not a believer in every aspect of the Labour Party, but I vote Labour because I am sure that party is much nearer to what I want than any other, and I cannot understand how anyone hopes to revive Liberal principles by keeping in a Tory government.

A third point raised was disagreement with much that appears in this *Journal*, particularly about Russia. I don't believe Russia is a paradise on earth, but I know that Czarist Russia was worse than anything that now exists. I dislike, as do all Englishmen, religious persecution, yet it must be admitted that the Orthodox Church of Russia has suffered far more for its politics than for its religion. I hope that the changes now taking place in that country will allow greater liberty of conscience. False religion may be destroyed by persecution, but not the true stuff, and religion in Russia may come out better in the end for its trials, if it be the right sort. Anyhow, in view of the poisoned prejudice that the capitalist press poor out about the Soviet, it is something to hear another side. The wise man hears both sides and makes his own judgement.

The last point was that my friend believed in "good relations with the Company." So do I. I don't believe in industrial war any more than international war. But I believe that a strong union, wisely led by sensible men, will in the end promote a better understanding between the men and the companies than will non-unionism. There are locomotive superintendents who are good straight men who do their best for enginemen. There are also others. The men cannot be left to the chance that gives one sort of boss or the other.

Were I an engineman I should be a unionist, not because I necessarily agree with all that the union officials think or do, but because I should want to stand by my mates and to help for the common good of the whole. I only wish all the "nons", many of whom are good fellows in their way, would think a bit more widely and act a bit more socially. Then they, and the rest of us, would be better.

EDITOR'S FOOTNOTE

In a 1930s context, guarded sympathy towards Communist Russia was not unusual. To my knowledge, ESW was not a political activist. Why he felt the need to air these issues is a mute question.

'I once congratulated a driver on punctual arrival at Liverpool Street. He replied: "I'm 20 seconds late." I said: "Well, surely, you don't count that?" His answer was: "Time's time, and 20 seconds late is not time." Either that brother was an inveterate leg-puller or the most meticulous timekeeper I have known in a lifelong experience of train timing'

February 1940

'This is a civilians' war now just as much as a soldiers' and in the front line of those who are winning it are those who are beating Hitler by keeping the wheels going'

October 1940

'I do not regard the railway director as a heartless plutocrat, but far too many of these gentlemen have no special knowledge of the railways, and no qualifications other than their social and business standing'

February 1941

'Railwaymen are reasonable, quite willing to take the good with the bad, so long as they know that they get a fair share of both the ups and downs, and are not made to suffer during bad times while having to struggle and threaten to get a share of better times'

February 1941

Behind three Kings and a Castle

January 1939

A few recent trips have afforded examples of the working of the powerful King class on the GWR. The class embraces 30 engines, 6000-6029, and they work from London to Wolverhampton and from London to Plymouth, either via Bristol or Westbury.

No 6008 brought me back from Wolverhampton with Driver E.H.Luscombe and Fireman J.Thomas, of that place, in charge. The load was only 270 tons, and after stopping at Birmingham and Leamington we left with 92 minutes allowed for the remaining 87¼ miles. Bicester, 33¾ miles, was passed in 35 minutes, and then we were brought to a dead stop at Blackthorn; but so smartly did we get away that the 9¼ miles from passing Bicester to passing Ashendon Junction actually took only 11 minutes, including the momentary stop. Even allowing for the light load, this was a very fine recovery. It was followed by a sharp spurt down the Chilterns, speed rising to the nineties, so that at Northolt Junction we were a minute to the good and, despite a couple of signals on after that, arrived exactly on time.

My next trip was with 6009, in the charge of Driver Poole and Fireman Shoote of Old Oak Common shed, on the 1.40 pm to Newton Abbott. The load was 310 tons at the start. We reached Reading, our first stop, 36 miles in 40 minutes after signals outside, and left with 61 minutes for the 59½ miles to our next stop, Westbury, an allowance exceeded by only two minutes despite a 15 mph slack for a bridge at Calcott and a pw outside Westbury. The 47½ miles to the next stop at Taunton were done in 47 minutes against a booking of 48, but one coach had been detached at Westbury. Taunton was reached to time, where another coach was dropped, and with the lightened load the allowance of 38 minutes to Exeter, 28 miles including the climb up to Whiteball, was cut by two minutes, our arrival at Exeter, then Newton Abbott, being a little ahead of schedule.

For the return journey on the Torbay Express, Driver Graydon and Fireman Southcombe of Newton Abbott had 359 tons behind 6023, and we left Exeter with 175 minutes for the 173½ miles knowing that a pw at Heywood Road Junction and the bridge at Calcott would take their toll. In consequence, Driver Graydon prudently got a bit ahead, passing Taunton in 33 minutes and doing the 27½ miles to Castle Cary in 24 minutes, so that although the pw check cost nearly five minutes we passed Newbury to time. But here fog took a hand and, as the distant was on before the bridge at Calcott and the home signal invisible, a very cautious approach had to be made to the check, and another five minutes went. Yet we passed Reading, 36 miles out, with 34 minutes for punctual arrival and should certainly have been in to time had not fog and a succession of signal checks spoilt everything after Reading. Yet the net time for 137½ miles from Exeter to Reading was actually only 131 minutes. Driver Graydon will have retired by the time these lines are in print. He is a keen member of our Society. I congratulate him on the excellent trip he gave me, and trust his retirement may be a long and happy one.

Another run was with Castle class 5035. Driver Hayward and Fireman Sharman of Wolverhampton were in charge. The load was 332 tons and Leamington was reached comfortably, two minutes ahead of time. From there to Birmingham time was kept despite a pw check which necessitated a recovery, and Wolverhampton was reached slightly before schedule. The Castle, on this showing, is fully equal to the work.

In all these runs not a minute was lost, but many gained, and the only late arrival was through fog; so that GW men who claim I have not done justice to their work must take this as some sort of compensation. Moreover, the GW have the prettiest and neatest engines in the country, I admit.

EDITOR'S FOOTNOTE

Catching up with two of the company's best-known engines, ESW's descriptions led to a productive exchange with Fireman Hatcliffe of Bath Road shed, Bristol (see following pages).

Learning about our own: a GWR fireman pitches in

January/June 1939

This first batch of letters found in ESW's archive is a detailed exchange with Fireman J.W.Hatcliffe of Bath Road depot in Bristol, occasioned by the January 1939 Journal *article. Although ESW didn't keep a record of postcards he sent in reply to letters, Fireman Hatcliffe's responses indicate a strong relationship, quickly established.*

Both sides were keen to meet but there is no evidence they did. At this range, what's extraordinary is that such an intelligent individual, who had been firing for 20 years before getting the chance to become a driver, had lost none of his pride in the job.

Paddington-Didcot running table (see letter opposite)

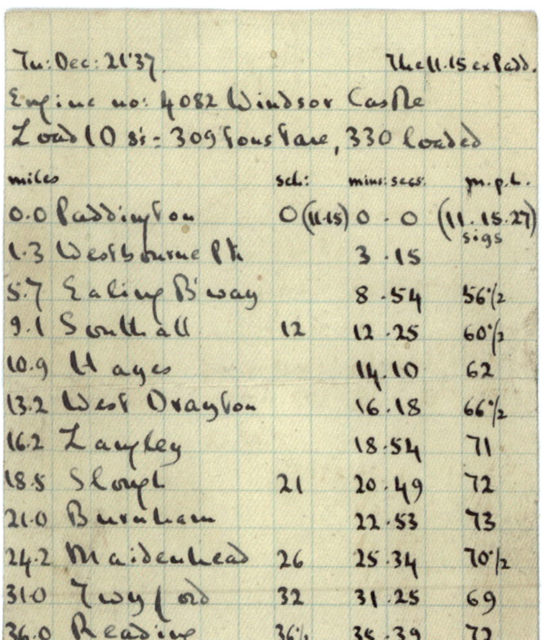

Tu: Dec: 21'37. The 11.15 ex Padd.
Engine no: 4082 Windsor Castle
Load 10 8's - 309 tons tare, 330 loaded

miles		sch:	mins: secs:	m.p.h.
0.0	Paddington	0 (11.15)	0 . 0	(11.15.27)
1.3	Westbourne Pk		3 . 15	sigs
5.7	Ealing B'way		8 . 54	56½
9.1	Southall	12	12 . 25	60½
10.9	Hayes		14 . 10	62
13.2	West Drayton		16 . 18	66½
16.2	Langley		18 . 54	71
18.5	Slough	21	20 . 49	72
21.0	Burnham		22 . 53	73
24.2	Maidenhead	26	25 . 34	70½
31.0	Twyford	32	31 . 25	69
36.0	Reading	36½	35 . 39	72

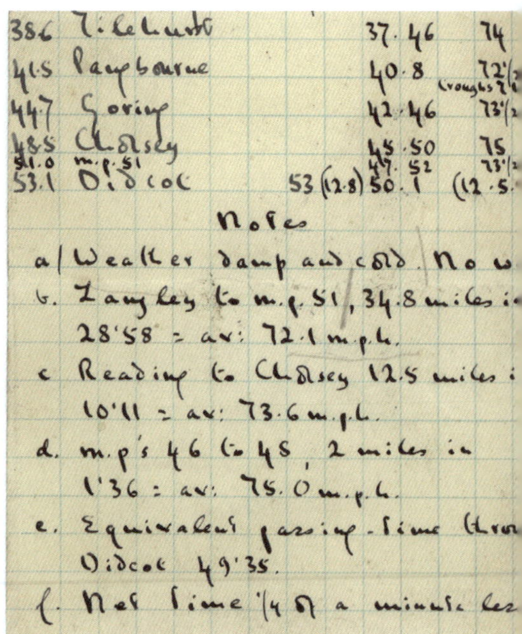

38.6	Tilehurst		37 . 46	74
41.5	Pangbourne		40 . 8	72½ troughs
44.7	Goring		42 . 46	73½
48.5	Cholsey		45 . 50	75
51.0	m.p. 51		47 . 52	73½
53.1	Didcot	53 (12.8)	50 . 1	(12 . 5

Notes

a/ Weather damp and cold. No w
b. Langley to m.p. 51, 34.8 miles i
28'58 = av: 72.1 m.p.h.
c Reading to Cholsey 12.5 miles
10'11 = av: 73.6 m.p.h.
d. m.p's 46 to 48, 2 miles in
1'36 = av: 75.0 m.p.h.
e. Equivalent passing-time thro
Didcot 49'35.
f. Net time ¼ of a minute les

36 Daventry Road
Knowle Ext
Bristol 4

January 10 1939

Dear Sir

I have read your article in the January 1939 issue. It was very interesting for me to read this all-GWR article and I can assure you that many more GWR footplates will have enjoyed the change. In the past I have been interested in knowing what other Rly Cos' engines can do. Now this article of yours has enlightened me in what some of our own men can do.

I have often thought that GWR men could do it as well with an engine in good condition and especially with good coal. We must of course admit that the main thing is a driver with experience in running fast trains. I have seen our speedometer showing 97 mph. The distant signal was at danger that moment, and my mate brought the train up to the home signal as cool as if we were only doing 20 mph. Of course it was a falling gradient and a light load.

It's the Three Cs – Cool, Calm and Collected – that make an express driver. Technical knowledge alone won't do crack runs.

Perhaps you will be interested to read the enclosed running table of one of our Bristol trains. This was taken by a passenger in the slip to Didcot and sent on to us. I think he was an official's son. On this occasion with a 40XX we were doing 93 mph through Dauntsey. If the GWR 40XX were kept in good order and a Castle cab put on (for a little comfort in working) they would do any Bristol runs and are very economical, but I'm sorry to say that I have just heard that we are to be a Castle depot at Bath Road. The Castles seem to be a very powerful engine, but they don't have such free and easy running.

Of course I am only writing with the experience of a 20-year fireman and just over a year on mileage runs. Will close now. Eagerly looking forward to your next article.

Yours truly

Fireman J.W.Hatcliffe
Bath Road Loco

'Bristol-Shrewsbury is a wonderful run for experience on gradients'

January 20 1939

Dear Sir

Yours to hand of the 16th inst and thanking you for your most appreciated reply. It is with regret that I shall not be able to see you as suggested on Feb 4. As my turns are at present, I will be on the 12.20pm Bristol-Shrewsbury that day.

I would have esteemed it an honour to invite you to lunch on that date and am certain I would have enjoyed a talk with you.

This Shrewsbury is a very difficult run for us as we generally have 14 coaches. I have known us to arrive five minutes before time. Last time, I noticed my driver going up over Langley Angle, with 1 in 40, on the 20 per cent without altering the lever. Of course I do not think all drivers do this. It is a wonderful run for experience on gradients.

That was a very smart run you mentioned on the LMS (Birmingham-London) especially with a stop and checks. I am looking forward to reading that article next month.

It is surprising the loss of time due to checks and stops. I have known us to go ten miles before we get back to over 60 mph with a Castle. When they are doing 65 mph and over they seem to be a different engine and ride well. Of course they are better balanced with four cylinders when running fast.

You mentioned the 29 class. Yes, they have been wonderful engines for fast running, also very easy for firing. Like the 40, I think if they had stronger framing they would pay for the extra expense by economising on coal.

Today I had 6009 King class on the 4.15pm from Padd-Bristol. We had 11 coaches, four stops and a check and were only five minutes late in. That was for time lost at stations.

My next turn for the 11.15am Padd-Bristol will be in seven weeks time, the first three days. That is if I have not passed for driving by then, as I am fifth at Bristol. We don't have anyone taking the examination at present.

Trusting I have not encroached too much on your valuable time.

Best wishes
Yours sincerely
J.W.Hatcliffe

‘Lately we’ve had a couple of hours at Paddington owing to the signal box being burnt’

January 27 1939

Dear Sir

I received your PC of the 21st inst. Thanking you very much. In my last letter I mentioned Langley Angle. It is a Welsh place and should have been spelt Llanvihangel.

As regards seeing you I appreciate your invitation very much but my time in London is not sufficient. Lately we’ve had a couple of hours at Paddington, owing to the signal box being burnt. Should ordinary engine working commence next week, on Mon & Tues I will arrive at Padd 1.45pm then wait about ten minutes for coaches to be taken off and take our engine to turn etc, returning to Padd about 3pm in time to work the 3.15pm back to Bristol.

Enclosed are the running times I took on Friday January 20 when I sent your last letter. We have no King class engine stationed at Bristol, we only had them for temporary working. I don’t regret it as we could have done that run much easier with another class. I am now going to get ready to leave with the 5.15pm to Padd.

Best wishes

Yours sincerely

J.W.Hatcliffe

4/15. EX PADDINGTON.- BRISTOL. 6/39. JAN20/ 39.

ENGINE. 6015. KING CLASS. 9.COACHES. 294,TONS.

DRIVER.F.ELLIS. BATH Rd. BRISTOL.

FIREMAN. J.W.HATCLIFFE. "

~~STOPPING AT READING. SWINDON.~~ CHIPPENHAM. BATH. BRISTOL.

MILES.	CHNS.				
		PADDINGTON.			
24	2	MAIDENHEAD.	PASSED IN	25Mins.	
36	PADD -	READING.	STOPPED "	37."	
17		READING TO DIDCOT. (CHECKED).	PASSED	17."	
		HIGHWORTH JCT.(CHECKED)			
41.	15.	READING TO ~~XXXXX~~.SWINDON.	STOPPED "	40."	Schld
16.	53.	SWINDON TO CHIPPENHAM.	" "	17."	18 mins
12.	73.	CHIPPENHAM TO BATH.	" "	14."	14 "
11.	38.	BATH TO BRISTOL.	" "	12."	15 "

80.Mph. through BOX.

Arrived in BRISTOL 6/40. 1Min. late.

No station time checked.

J.W.H.

SAT28. JAN. 39.   ENGINE. 4096. HIGHCLERE CASTLE.

5/15.   5/15. _BRISTOL. - PADDINGTON. 7/15.

BRISTOL. DEP. 5/17 .2MINS LATE. LOAD 10.= 337.TONS.
COALPIT HEATH.5/32. STOPPED (TIME INTERVAL WORKING) VERBAL.
CHIPPING SODBURY.   CHECKED.
SWINDON. SLIP" 2.   (8.= 289. TONS.)
READING.   "   1.   (7.=   223.   ")   25 MPH.
PADDINGTON. ARR. 7/18.   3.MINS LATE.

MON"30. JAN. 39.   ENGINE 4096. HIGHCLERE CASTLE.

		SCHLD	PASSED.
BRISTOL. DEP. 11.45. LOAD 9= 295 TONS.		11/45.	
WOOTTON BASSETT.		12/26.	12.25
RUSHEY PLATT.	CHECKED. down to 20 mph		
SWINDON.		12/32.	12/30.
STEVENTON.		12/50.	12/48.
CHOLSEY&MOUL;	CHECKED. " " 15 mph.		
PADDINGTON.		1/45.	1/42.

ARR. 3MINS. EARLY.

TUES.31. JAN. 39.   ENGINE. 5070.

	SCHLD.	PASSED.
BRISTOL. DEP. 11/45. LOAD.9= 295.		
WOOTTON BASSETT.	12/26	12/26.
~~SWINDON.~~		
~~PADDINGTON.~~		
SWINDON.	12/32.	12/31.
STEVENTON	12/50.	12/50.
PADDINGTON. ARR. 3.MINS. EARLY.	1/45.	1/42.

Fri 3rd Feb, 39.

Dear Sir.

I thought you may like to see the running of these three trains
On the Saturday we only lost 1 min with a stop & check.
The Monday with same engine 2 checks and 3 mins early (a different train)
Tuesday same train with different engine, a through run 3 mins early.
The free running of 4096 was very good on 15%. The 5070 was
mostly on 17%.

Yours Sincerely
J. W. Hatcliffe.
56, Daventry Rd Knowle Extn. Bristol 4.

'My time for firing is nearly over... I hope that you will ride with us before I leave here'

June 19 1939

Dear Sir

I received your welcome PC of June 4. Thanking you for same. I sincerely hope you will find it convenient to ride with us on one of the first three days next week. I have told my driver of your intentions; he said if only he could have an engine like the Coronation he would give you a good run. Considering the load we've I think he does as well as any Bristol driver.

I noted in your article with reference to the trip you had with Driver Fisher that coming through Box Tunnel the wheels picked up. Others have had that experience. When we enter the tunnel I notice that my mate shuts the regulator hard then opens into the first valve. If the regulator is shut down gently, sometimes it is doubtful that it has gone into the first valve – and that is what I thought caused the slipping, although I have not had the engine slip when I`ve been through there.

That run from King's Cross to York in this month`s *Journal* is wonderful; 188.2 miles at an average of 71.9 mph.

Sorry that I did not get a chance to call on you when I went to Chingford for a weekend. I had another relative to see at Ilford and the time simply flew by, especially taking my two sons round the City.

It seems that my time for firing is nearly over now. We have been passing one a week from Bristol lately and I am the third for the exam. Possibly I will be moved from here when I am made a driver as we get sent where the vacancy occurs. I have been firing 20 years last March so will welcome the change.

Trusting I have not bored you too much with my letter and that you will ride with us before I leave here.

Best wishes
Yours sincerely
Fireman J.W.Hatcliffe

Of drivers and enginemen

February 1940

Anyone with ordinary common sense can learn to drive a car. Some, indeed, learn without any sense. But not one motorist in 50 is more than a driver. I don't suppose more than ten per cent of them could even do a simple repair. And as for nursing and studying their engine, that seems altogether beyond the average motorist. Thousands of motorists cannot even keep an eye on their petrol and oil gauges or ammeters, fail to put back their trafficators, and run past lights at red.

I suppose that on an engine these folks would forget the blower, the sanders, burn the boiler bottom, and take the signals on the wrong road. All of which is to indicate that a driver need not be an engineman. But on every line there are drivers and enginemen. The engineman is a mechanical enthusiast. He not only knows his engine, but he knows the why and wherefore of its working. He takes a pride in getting the best out of his machine, and is willing to study not only the peculiar features of each type but to find out the whims of individual engines. One, however, should add to this the proviso – if he gets the same engine often enough.

For example, there were men who soon mastered the art of getting the best work out of the Midland compounds when that type became a standard class on the LMS. Indeed, some Scottish drivers in particular seemed to me to get more out of the compounds than I thought the engines were capable of doing. Other men who were good enough drivers never liked the compounds, and never managed them properly.

The same is true of the Capriotti-valved engines, and of the Sandringham class of the LNER. The modern long-valve travel engines can be notched up as far as ten per cent, yet I know men who have always been used to driving on a small regulator opening, and 20 to 25 per cent cut-off, who seem quite unable to adapt themselves to any other way of working.

On the road the engineman not only knows the timings of the working timetable but his own variations of them, for in certain cases a minute or two before or after the booked times is better practice at some timing points. Usually, however, the engineman likes to have a couple of minutes in hand, and when he knows a pw delay is ahead, gets a bit more in hand to meet the delay in advance and keep time despite it. He knows what's ahead. It is no use punching one's engine when a train ahead of you is due in only five minutes before you, but some men never trouble about that and get stopped by signals as a consequence. The engineman does not rely on getting a minute or two back by exceeding the limit where there is a speeding restriction. Some of the fastest drivers are the most cautious in observing limits.

I have known a man show his good sense by dropping time. Newbold troughs were out of action, and a heavy train had to go from Hademore to Castlethorpe, about 55 miles, without a drink. The driver wisely eased it, losing a few minutes, but not nearly as many as would have been lost if a stop had been made at Rugby for water.

Last summer I came up from York with a load of 575 tons and a mile-a-minute timing. A fellow passenger was anxious to arrive on time, as he had hopes of getting a connection on from Waterloo which could just be caught if we got in to the minute. I told him to make his mind easy. We should be there if we had anything like a clear road, I said, and my confidence was confirmed with two minutes to spare. I knew that Driver – no, I will give no names, but there is no better man on the road, and few as good – was in charge.

You won't believe my concluding story. I can't help it if you don't. It is true, all the same. I once congratulated a driver on punctual arrival at Liverpool Street. He replied: "I'm 20 seconds late." I said: "Well, surely you don't count that?" His answer was: "Time's time, and 20 seconds late is not time." Either that brother was an inveterate leg-puller, or the most meticulous timekeeper I have known in a lifelong experience of train timing.

EDITOR'S FOOTNOTE

ESW digs into his records to illustrate a belief that some train drivers are more worthy of their craft than others. We don't know how this went down with ASLEF's driver brethren.

Wartime speed limits unfair to some

May 1940

Although there is a general speed limit of 60mph now, and the quickest timings are not supposed to exceed 50 miles an hour from point to point, there is an extraordinary difference in what is expected between these limits.

For example, Driver Williams and Fireman Trevis of Salisbury, on the 2.50pm from Waterloo, with King Arthur class 457, had a load of over 400 tons. The distance, 83¾ miles, is booked in 112 minutes, including three stops optimistically booked at one minute each; actually they occupied over seven minutes. In addition, we had a nasty pw delay costing five minutes. If we allow four minutes each for the stopping and restarting, which is not too much by any means, that makes in all 24 minutes to be deducted from the 112. Actually, we arrived eight minutes late, which means we slightly improved on the difficult schedule, but it wants doing when, with such a load and timing, stops occur every 25 miles or so.

In comparison, take a return from Birmingham with Royal Scot 6140, Driver Scroggins and Fireman Ball of Camden. The engine is much more powerful than the King Arthur class and the load was less, 354 tons. From Rugby, slightly less distance from London than Salisbury, ten minutes more is allowed, which is the amount of time booked at Northampton and Watford, the two stops. Of course, Driver Scroggins fulfilled my promise to an anxious fellow passenger who had a connection to catch. I told him as we left Birmingham that if we had the road I'd guarantee that he'd get his connection easily. Despite a few delays we had time in hand, and arrived two minutes ahead, but of course the work was much easier.

Again, 73 minutes for the 53½ miles from Paddington to Bicester sounds easy, but the new Tube extensions cause severe pw delays, and additionally we had signals at Saunderson, so that Driver Harry Hunter and Fireman Weston of Wolverhampton had to work King George 11 hard to keep to time, though the load was not more than 347 tons. The 14 miles on to Banbury is booked start-to-stop in 17 minutes, a very hard task under present limits, but accomplished exactly. Mr Antrobus, a regular traveller on this line, well-known to drivers, was with me on this trip and regarded the effort as entirely satisfactory.

The return was made with a Castle class engine, 5067, and Driver Cowen of Old Oak Common shed, the load being 13 bogies, and was skilfully done with a number of irritating delays. Bro Cowen prefers a Hall class with 6ft wheels for such trains, since it is well up to the maximum speeds allowed and better for climbing and starting.

I had a couple of runs from Grantham to King's Cross, non-stop, 105½ miles in 128 minutes. On the first Driver Dodd and Fireman Bellamy, of Grantham, had a bad trip with a V2 2-6-2, 4818, which normally would not have had any difficulty taking the 14-coach train. But another wartime handicap was against us. She wouldn't steam, and we dropped ten minutes in consequence. But the second trip, made with the same crew, was with a streamlined Pacific, 4903, and a huge load of 18 coaches, giving a gross tonnage of about 600. Yet the climb to Stoke summit and the run down to Peterborough was so smartly done that we passed the latter place, 29¼ miles, in 35 minutes. From Connington to Sandy we met signals all the way, but after that ran on, doing the last 44 miles in 52 minutes, a capital bit of work with such a load.

A trip on the 1.25pm to Doncaster, as far as Grantham a 15-coach train with a similar engine, 4487, was easily done by Driver Hoskins and Fireman Cooke, of King's Cross, with time in hand all the way, whilst Driver Thompson of Grantham, who took us on from there with a 2-6-2, 4844, did equally well to Doncaster.

I also have a good account of an LMS trip to Nottingham with Stanier 3-cylinder 5616, Driver Betts and Fireman McKerran of Kentish Town, and 11 coaches. This is a run that makes five stops and the 123½ miles is not easily timed at 177 minutes if these be considered. We were delayed by signals and excess time at stations to the extent of 18 minutes, but were only five minutes late, and this was done without any excessive speed at any point by good work in getting the engine going and keeping her at it.

EDITOR'S FOOTNOTE

How drivers worked to keep trains on schedule during the phoney war months, although limited to 60mph to save on coal.

Where stops make timekeeping impossible

June 1940

As mentioned last month, war schedules call for point-to-point speeds not exceeding 50mph and maximum speeds of not over 60mph. Given a long run that presents no great trouble – even with a big load. But put in a number of intermediate stops, with close-cut station times, and trains are bound to be late. The starting and stopping of a heavy load makes it difficult to keep point-to-point times and any time lost at stations or by checks is unrecoverable.

I noted a run with the 2.50pm to Salisbury last month. Almost exactly the same experience befell me. In this case King Arthur class 454 was hauling 14 bogies and was handled by Driver Galpin and Fireman Cox of Salisbury: 83¾ miles booked in 112 minutes, with three stops. Excess station time and a pw check cost eight minutes – the amount of our lateness – which shows that under such conditions it is hard to keep schedule and impossible to recover time lost.

Or take the 8.30am from St Pancras to Derby with 3-cylinder 4-6-0 5565, an engine technically more powerful though with less heating surface than the King Arthurs. In the 130 miles (via Trent) there are no less than eight stops, and the longest non-stop run is 30 miles. The load was 393 tons, overtime at stations 13 minutes, and ten late in. Driver B Clarke and Fireman Duckworth of Trafford Park worked the engine hard, too.

I timed a similar engine (5588) with a bigger load – 447 tons – from London to Preston, but here we had only one stop, Rugby, between London and Crewe. Driver Wilson and Fireman Cartmel of Blackpool worked the train skilfully, and the time from Rugby to Crewe was only 84½ minutes for 75½ miles which, with such a load, would have been good even with prewar timings. On to Preston, with stops at Warrington and Wigan, as well as the usual speed restrictions on that stretch, the task was harder but was done with a couple of minutes to spare. This was an excellent trip, on which I congratulated our members, but it also shows the difference between running under such conditions and running like that in the trips above-mentioned.

The return, made via Stoke with 5568, Driver Lea and Fireman Abercrombie of Longsight and a load of 444 tons, provided a non-stop run of 146 miles in 173 minutes, the net time after allowing for checks being about five minutes or less.

I must not omit the return trip from Salisbury as it was made with a King Arthur, 755, recently fitted with a Lemaitre five-way blastpipe like the Lord Nelsons have, and a "washtub" chimney – together with an increase of cylinder diameter from 20½ to 22in, and streamlined ports. Driver Lush and Firemen Hone of Nine Elms worked a 390-ton train very successfully. The run from Andover to Waterloo – 66½ miles in 76 minutes – is the fastest trip in England under war schedules. Actually it took 82 minutes, but as this included a dead stop for four minutes at Whitchurch for goods ahead, and another stop outside Waterloo, the net time must have improved well on schedule. The engine was reported to be an unusually free steamer but a little heavier on coal than the standard type. Her exhaust is very soft and her hill-climbing excellent.

Another excellent run was provided by 802 from Charing Cross to Ashford, tightly timed in prewar days and which now, with an extra two-minute stop at Tonbridge, only gets ten minutes more. The load was 230 tons and Driver Barker with Fireman Boxall, members of our Bricklayers Arms branch, did fine work in arriving to time after signals at London Bridge costing two minutes and two pw checks. The 26½ miles from Tonbridge was done in 31 minutes, start to stop, with a pw delay of three minutes – very smart work, even allowing for a light load and a level road.

Greetings to Driver Reg Churchill and Fireman Wooldridge of Eastleigh, who gave me an interesting trip with a Drummond tank; to Driver A Swann of Boston and Motorman Reg Wood of Bury, companion passengers with me; to Driver Lee and Fireman Weston of Bath for a run on the old Somerset and Dorset; to Driver Chorwell of Exeter and Fireman Ganderton of Worcester. Also to Messrs Marshall, Hayden and Kiddle of Salisbury and another member from that city whose unsigned message I acknowledge gratefully.

EDITOR'S FOOTNOTE

An opportunity for ESW to catch up with members across a spectrum of routes, and to note refinements to a King Arthur locomotive.

A strange request and other tall stories

October 1940

Adolf's air raids, during one of which a very large souvenir was left within 50 yards of my front door, have curtailed my travels. The gentleman who dropped the souvenir did not return to Germany, I am glad to say. It made me admire all the more the bravery of enginemen who carry on. The bus drivers can stop their vehicles and hop to a shelter. The railmen just go on, and one of the first line factors of winning the war is British railways. Here's to the men whose weapon is a spanner, and whose uniform just a "tin hat" and blue overalls. God bless you, gentlemen!

Here, by the way, is a story to which I can't give an explanation. Perhaps some reader would like to solve the mystery. One of our mainline drivers told me, before the war, that he was approached by a stranger who said that he understood that if the mainline train were ten minutes late a connection at a certain junction would not wait for it. The driver said that, as far as he knew, that was so, but there was another connection not long after, so the delay would not be serious.

Whereupon the stranger told the driver that there would be £5 for him if he arranged matters so that the connection should be missed. The astonished driver asked what was the game and was told that it was no business of his, but the money would be sure if he held his tongue and arrived late. Not liking the business, the driver declined. The man then said that the communication cord would be pulled and the train delayed in that way. The driver replied that in such a case he would certainly point out the offender. Whereupon the man walked off, nothing happened, and the train arrived at the junction to time. The mystery is why anyone would make such a request. Was it to hold up some commercial rival from submitting a contract, or what?

More understandable is this strange coincidence. An airman who had overstayed his leave approached another driver, who shall be nameless, and said that he would be obliged to pull the chord and stop the train at the station near his aerodrome and though he supposed it would cost him £5 it was better than being court marshalled. But, he added, suppose close by there you felt something wrong with the engine, stopped to examine, found it right and went on, wouldn't that be odd? It is an amazing coincidence that just at that point the driver did not like the sound of things, stopped to investigate a knock, found everything all right, restarted and regained the lost time. I asked him if he had looked in his pocket and he replied that some people asked far too many questions!

An extraordinary adventure happened a good many years ago to the train crew of a Leeds-Sheffield train. Hearing a sound behind, the driver turned to see a man drop onto the tender from the roof of the first coach. Seizing the shovel, the fireman threatened the intruder, who promptly surrendered. His story was that he had quarrelled with his girl, who taunted him with being a coward. To prove his valour he climbed out onto the footboard, then to the roof, and dropped onto the engine with a wild idea of stopping the train. Had he not done so at that minute, a tunnel would have swept him off and killed him.

Another story I find hard to believe, though a locomotive inspector told me, is that a Southern driver, having shunted off a van at Winchester, returned to the train but for some reason was not coupled on. The guard gave the right away and he ran to Basingstoke without his train and without knowing it.

Many years ago a LNW driver who was a bit of a dog fancier saw a Yorkshire terrier apparently lost and wandering about Euston. He slipped it into his tool box and drove off to Rugby. As he entered the station he saw a policeman waiting at the platform end. The fireman took the dog under his coat and leaned over on his side. The policeman climbed up on the engine and was making enquiries when, as his back was turned, the wily driver broke the gauge glass. The policeman leapt for his life, terrified, and the fireman dropped down and set free the dog in the station, where he was subsequently found. As no proof was to be had in the manner of his getting there the matter was dropped.

Yes, some odd things happen on the line.

EDITOR'S FOOTNOTE

Written as the Battle of Britain morphed into the Blitz; after a rousing salute to the bravery of enginemen, ESW spins a few yarns to cheer readers up.

Enginemen on the Front Line

November 1940

I am not amazed at the way enginemen are carrying on, though I am proud of them. They are on the Front Line. A newspaper account spoke of a shunting engine moving up and down the yard whilst 60 bombs fell, and greeting each of them with a blast of the whistle! Other men have been machine-gunned as their trains were running or whilst in stations. There is a growing casualty list. But they stick it.

A word for the wives too. It isn't easy for a woman to wait at home when her man is away, and she has no idea when he will be home again, or where or how he is. But they stick it too. God bless 'em. This is a civilians' war now just as much as a soldiers' and in the front line of those who are winning it are those who are beating Hitler by keeping the wheels going.

During the last war I noticed that soldiers were far less bitter against the enemy than civilians. I was quite able to appreciate the attitude of a driver I met the other day who said he was sorry for the enginemen in the much-bombed Ruhr railway yard of Hamm. After all, when one has been through the horrors of war oneself it is easy to feel a bit sympathetic for the other fellow, even if he is on the other side, who is getting it too.

I knew a few German enginemen before the war, and they were very gratified that I, as an Englishmen, showed interest in their engines, and as far as my bad German allowed we carried on a conversation. I found them decent enough fellows. Now that night by night our bombers are smashing up communications, some of my former acquaintances must be getting a hot time of it in the goods yards.

Amongst the differences I noticed between British and German railways was that the Germans had more provided for them. Each carried a good watch, supplied and kept in order for him. More important still was the careful instruction given as to working the various types of engines. We are amazingly haphazard here in this respect.

When, after the 1921 amalgamation, engines from the various companies were being mixed up, in many cases no attempt was made to show the drivers and firemen the best ways of working them.

For example, a Midland compound has three cylinders, two low pressure and one high. On opening the regulator the engine works as an ordinary simple, steam being admitted direct to the low pressure cylinders. By putting the regulator full over, a slide cuts off the direct admission to the low pressure cylinders. The main ports to the HP chest are opened, and compound working begins. If, when running, the regulator is closed and then opened to the auxiliary steam position used when starting as a simple, a semi-compound working is secured.

All this was familiar to old Midland men, but when a former LNWR man I know found himself booked for a compound he asked how it worked. The reply was, "Find out." Is that fair to man or engine?

There are many gadgets on modern locomotives which are new to many enginemen, various types of feed-heaters, sand-guns, new types of lubricators, exhaust steam injectors and pumps, snifting and by-pass valves and special gears such as the Caprotti.

In days when a man may get any sort of engine given him at a minute's notice, it would surely be good business to issue an instruction book giving particulars of the various peculiarities, and suggesting the best regulator and cut-off positions for normal working.

A few enthusiasts will tell me they prefer to find out things for themselves, but even so I think it would be better to let all know what is standard practice even if some should find it needful to modify standard methods because of the peculiar conditions of some of the roads traversed and turns taken.

The older type of engine could be thrashed, as were the LNW Georges, which at times were loaded up to 400 tons. The modern engine is not designed for such handling, and to get the best of its possibilities needs skill and thought. Surely it would make for economy and efficiency if more care were taken to explain the designer's ideas to the men who are responsible for carrying them out.

EDITOR'S FOOTNOTE

Written when Nazi invasion was still a possibility. How many readers shared ESW's sympathy for German enginemen bombed by the RAF?

The prowess of Midland singles

January 1941

I have been turning up an old pocket book which records some of the first railway journeys I timed. I was a young man then and records were not made with the care I now employ. But, as the earliest of them goes back to the beginning of the century, they indicate in an interesting way how we have progressed.

At that time the railways were in a state when increasing loads, due to heavier coaches, corridors and dining cars, were being hauled by underpowered single-wheelers and 4-4-0s. There were many drivers who performed wonders with them, aided by that forbidden instrument known as James, jemmy or jimmy. If any of my younger readers do not know what they were, let him ask one of the old brigade.

But there were still lingering traces of the pernicious practice of allowing a bonus for coal saved, which led to bribery at the coal stages where labourers loaded coal on the tenders, and to "coal-dodging" which meant that not a single minute was made up and full advantage taken to claim every possible minute and a few more due to delays. I notice that fewer of the journeys then were on time. My records don't show much time booked against the engines, but not much made up either.

Trips on the Great Northern include several with the famous 8ft singles and others with the Stirling 2-4-0 class on the mainline, and on the Grantham to Sheffield line the then-new engines, still running and now numbered 4342 and 4346. I can find nothing specially worth noting about these runs. No 93, having run from Grantham to Peterborough start-to-stop in 32 minutes, evidently found the load too much and came off with a hot box.

Most of my trips were then to Sheffield, and often by the Great Central route. Loads were light, often only four bogies, but the engines used, a few of which still linger such as D6 class 5852, did not perform great feats, though I have one good trip down to the credit of Driver N Grain, who some of my readers will remember.

For a Midland run with the Scotch express, on a winter night of deep snow, Driver Barlow of Kentish Town left Leicester with a single 130 and reached St Pancras to time after a start 11 minutes late, doing 99 miles in 105 minutes. Another single, or spinner as they were sometimes called, 115, driven by Driver G Beddeson of Nottingham, with a light load, did the 60 miles from Nottingham to Wellingborough in 64 minutes, including a severe check at Melton. Restarting from Wellingborough, the engine made the climb over Sharnbrook so smartly that Bedford, 15½ miles, was passed in level time and the 65 miles to St Pancras took only 72 minutes, including two dead stops at signals at Harpenden, and another check almost to a stop at Hendon.

I should mention equally good runs by 116 and 1870, handled by Driver Grocock of Kentish Town, and 2604, worked by Driver Foulke of Derby. My experiences are not wide enough to make my opinion more than an opinion, but so far as they go they certainly indicate that the Midland singles were better engines and did better work than the more famous Great Northern class.

EDITOR'S FOOTNOTE

War curtailing journeys, ESW looks into his records of timed trips going right back to the 1900s, some of them surprisingly quick.

The future management of railways

February 1941

Will the end of the war see nationalisation of the railways? The last war was followed by grouping. That marked the abandonment of the old cut-throat competition era, and few would wish to see it return. Grouping is only a half measure. Unified control of all lines would greatly increase the advantages. Yet such control might bring with it disadvantages which would outweigh any gain the arrangement could offer.

Railways are a national necessity, far too important to be left in private hands. The community, as a whole, is too profoundly affected by the services they render and the charges they make to let boards of directors have a free hand. The control now exercised by the Ministry of Transport indicates that.

But is State ownership the solution? Is it wise to let the head of a state department be a politician, utterly unacquainted on the practical side with the job allotted to him, and to whom it is just a step in his career until he is appointed to some other post, rather than a life interest?

Is the efficiency or, I might add, the courtesy of the Post Office equal to that one receives from private firms? If, then, private ownership and ordinary State control both seem defective, what method can be suggested that will give better results?

First of all, I would like to see the shareholder eliminated. I do not desire to see anyone robbed; but most shareholders of ordinary shares, few of which pay any dividends, would gladly accept a comparatively small sum in exchange. The holders of the more profitable shares would receive State bonds, which would represent a fair equivalent.

The railways then would belong to the community as a whole, to be administered for the general good without thought of any one class. They would have to pay, of course, but the advantages of efficient unified management, the elimination of overlapping, and the coordination of road and rail transport, would provide for that.

Next comes the question of management. I do not regard the railway director as a heartless plutocrat, but far too many of those gentlemen have no special knowledge of the railways, and no qualifications other than their social and business standing. I would like to see general and sectional boards of management appointed by the chief interests concerned.

First, the State, responsible for the money which has been spent to purchase the railways. Its representatives would be men who understand the financial side of a great organisation. Next, the traders and passengers who use the railways. They should be represented by businessmen who know what the public needs. Last, but not least, adequate representation of the men who work the railways. Nothing could do more to secure goodwill and the absence of disputes and strikes than this.

Railwaymen are reasonable, quite willing to take the good with the bad, so long as they know they get a fair share of both the ups and downs, and are not being made to suffer during bad times while having to struggle and threaten to get a share of better times. Such management would in no way interfere with the unions. It would provide for their point of view being fully represented at all times.

In the old days the designer of locomotives was also the man responsible for dealing directly with enginemen. Because of the greater complexity of the big groups, now it is often the case that the superintendent of motive power is another individual. True, the ordinary driver cannot design a locomotive, but if the designer had a good talk with some of the best drivers before proceeding with his next design we would get better engines all the same.

Finally the State, representing the community as a whole, should retain the right to make, through suitable outside experts, occasional examination of the working of the railway system. This would be a means of guarding against any tendency to bureaucracy, slackness and getting into ruts. It is a method already employed by the inspectors of education in this country, and in this case it works well.

Railways run on some such lines as these might well be of far more benefit than they have ever been to the railwaymen and the country as a whole.

EDITOR'S FOOTNOTE

ESW was floating his ideas for postwar change at the very moment Britain faced the German Blitz. His suggestions are more on continental than British lines – three-way management between the State itself, the business/passenger community, and the railway unions. How much aggro could have been avoided with such an arrangement? Combining the private companies would, ESW assumed, offer the nation a financial return. In reality, postwar railways demanded huge investment they never received.

Early or late cut-off?

March 1941

I was interested to hear my friend, Driver Bishop of Camden, on the air the other night, and his story of the sailors who got on the tender at Crewe for a ride up. If a bit too much water had been taken at Whitmore troughs, the sailors might have had an experience they had heard of at sea but wouldn't expect on land – being washed overboard.

I have read with much interest the book of another Camden man, Driver L.A.Earl, called *Speeding North with the Royal Scot*, and congratulate him on it. I wish some more people would do the same thing. Mr Earl, an NUR man, reads these articles, and I have been carrying on correspondence with him on the cut-off question.

His own method is to start with the full 75 per cent, open the regulator slightly, and as soon as the engine begins to move, close the regulator and notch up to 15 per cent. As soon as he feels the train taut he gives full regulator and 25 per cent, always opening the sanders at starting. The idea of getting the train taut is clear, though some drivers do not favour it, believing that there is a strain on the engine induced by doing this. Mr Earl's authority is on the other side, and that is interesting. He tells me that on the Carlisle link with the Pacifics he never needs to use more than 25 per cent for timekeeping.

On the other hand, in the February *Locomotive* magazine, there is an article by Mr Nock, a well-known expert, who rode on the footplate of the 10am Euston to Rugby, the Royal Scot, on the prewar winter schedule of 56 mph average. The load was 475 tons and the engine 6111 of the Royal Scot class. The driver is not named but his method was unusual. He used 40 per cent cut-off and varying regulator openings to Tring summit, then 33 per cent and auxiliary port of the regulator for the rest of the trip. This was the only change all the way, and time was well kept. Coal consumption was given at 45lb per mile up to Tring. As it was not a test run I presume this is an estimate. It is a point of importance, however.

My own ideas have always been for early cut-off and using the expansive powers of steam. Some of our latest engines can be cut back to ten per cent, and the provision of long valve travel allows for this. But on the Southern, when I was on the footplate, the driver of a Lord Nelson never varied the cut-off of 20 per cent all the way to Salisbury, and we ran easily with no great amount of firing.

On the GW, the 100 miles from Badminton to London is often done with 15 per cent all the way, but it is a wonderfully level road. When I was last on the footplate over that stretch, an advance to 18 per cent enabled a Castle class engine to regain three minutes on a mile-a-minute schedule in about 30 miles. That sort of thing can't be done from Dumfries to Stranraer, Newton Abbott to Plymouth, Salisbury to Exeter and on similar stretches. But the point is that both the early and late cut-off methods seem to be justified by results even on the same engines.

EDITOR'S FOOTNOTE

ESW feels comfortable mentioning a book by, and his correspondence with, an NUR driver who reads Locomotive Journal. *See page 124 for later thoughts on cut-off.*

Responses to my February 1941 column

Last month's ideas about the future of the railways have already brought me some criticisms – mostly kindly. It will be understood that I was speaking for myself and wondering if there is any way to avoid private control and purely State control.

The parties might be equally represented on the boards. There should be a minimum wage for all, on a sufficient basis, and bonuses according to profits. As a rule overtime should not be encouraged. Where necessary it should be well paid for. Extra payment might be made for specially onerous and responsible jobs.

I am in favour of not allowing new entrants to come into the railway service unless they belong to a union and pay their subscriptions regularly. The political levy must, in fairness shall I say to the more foolish brethren, be optional.

A pension scheme for all should be part of the programme, and full provision made for the men to present any matter affecting them to the board of management. I am also in favour of the men electing their own inspectors. I believe that some such scheme as this would make for fair conditions and the absence of trouble.

The Channel Packet and a Castle baptism

April 1941

So the Southern have joined the company of streamliners. I haven't seen the new engine in the steel, shall we say, as it can hardly be "in the flesh", but it looks fine. If it does as well as Mr Bulleid's existing types are doing it will make history. I am conservative enough to prefer the appearance of the ordinary engine to anything streamlined, and I fancy the closed-in cab is more dusty than the open type. I hope the Channel Packet, as she is called, will not copy her namesakes and roll.

Now that the Southern have a Pacific class, the GWR, who built the first Pacific in the country, the Great Bear (now rebuilt as one of the Castle class) will be the only group without that type, and without full streamlining, since only a couple of engines were streamlined by the GWR, and that only partially.

Talking of the GWR reminds me that I initiated my little grandson David, a keen locomotive enthusiast at 3½, as footplate man on the GWR. I took him to Paddington to have a look round, and the first train we saw was in the charge of Driver Brayley of Bath Road shed, Bristol. He and Fireman Pett invited us up and David, rather scared at first, made his acquaintance with the footplate on Beverston Castle. Since then he has talked continually of it, of the big fire and lots of coal in the tender, usually adding "But I got my knee dirty". I record this to reassure readers that I am making preparations for another passenger to continue these articles when I have booked off from my last trip, and to remind my GWR friends who think wrongly that I am pro-LNER that my grandson begins as a GWR man!

I made a war journey to Yorkshire from King's Cross the other day. The trains are so lengthy that the engine is almost always beyond the platform, and it is difficult to get a word with the driver. We had an A2 Pacific, 2555, which worked through to Leeds. Driver Hall, of Grantham, started away with a 522-ton load and, despite a couple of bad checks, was stopped outside Peterborough well up to time. From Grantham, Driver Morris was in charge. I think he is usually on the Grantham Atlantic link but he maintained the good work and, though time was lost at stations, not a moment went against the locomotive all the way.

On the return journey I joined at Wakefield where a similar engine, 2547, Driver Kitchen of Doncaster in charge, came in with the enormous load of 21 coaches, 632 tons, and yet succeeded in keeping time to Doncaster, including a dead stop outside.

Driver Catchpole of that shed took on and, with stops at Retford and Grantham, succeeded, while gaining a minute here and losing it there, to keep exact time to Peterborough, though the delay of drawing up twice at each station and awaiting connections had made us 30 minutes late.

Here we changed engines and a 2-6-2 mixed-traffic engine, 4864, gave Driver Beards of Peterborough and his fireman a hard problem. She was not steaming well and it was a continuous struggle to keep up the boiler level. Consequently it was not until we were past Stevenage that we really were able to move a bit, and the last part of the run was the best. It was one of those cases in which the driver has to choose whether to struggle on though losing time or stop for assistance and perhaps lose as much by so doing. Driver Beards wisely stuck to it, and we got in no later than we would have done by stopping for a pilot.

I have read of some good work on heavy passenger expresses by these Green Arrow class engines, but my own experiences have not been in any way outstanding, and I find it difficult to believe that, despite their smaller wheels, they are anything like as good with bumper loads as the Pacifics. There are not many engines, even on war timings, would keep time with 632 tons of crowded train behind – indeed, if one reckons the gross load at about 700 tons. Yet the A2 Pacific, 15 years old, did so without great strain. Wartime travel is not very comfortable or punctual, but it teaches one something about loads.

EDITOR'S FOOTNOTE

Travel exigences proved no obstacle when ESW introduced his first grandson to the footplate, though David proved less entranced by steam than his younger brothers (both LNER men).

ESW commentary & letters in the *Journal*

Sir Nigel Gresley led the way

May/August/November 1941

In May 1941 ESW published the following short tribute to Sir Nigel Gresley, who died on April 5 that year:

We all regret to hear of Sir Nigel Gresley's death. As designer of the LNER engines since the last war he made his company second to none in locomotive work and pioneered the speed trains of the day in this country. His A1 class of Pacifics came out in 1922. Recently, by the way, I referred to them as A2 – why I don't know, for I am well aware that the A2 class were Sir Vincent Raven's design. The more powerful and streamlined engines were, after all, developments of Gresley's A1 and A3. I know of no other engines in this country with a pressure of only 180lb which haul such loads. They can still take 500 tons at a mile a minute start-to-stop over any stretch between London and Newcastle. Perhaps the secret of Sir Nigel's designs lay in generous heating surface and well-designed ports.

Then, in his column of August 1941 he wrote:

A letter from a Peterborough reader makes an interesting point that the LNER Gresley engines are not as good as I think them to be. He says that the old Atlantics of the GCR Director class of 4-4-0 engines would easily do the fast running of the Pacifics, and were more comfortable.

It cannot be suggested that they do so with the same loads, surely. Nor can I agree that they are capable of the same speeds. Pacific 4468, worked by Driver Duddington, holds the record of 126 mph, and 2509, worked by Driver Taylor, did 43 miles at an average of 100 mph with a 220 tons train. The valve gear and exhaust ports of the older engines would not permit such feats. The newer engines, in addition to such speeds, now haul loads at times exceeding 700 tons and can keep wartime schedules with them. These facts surely prove them to be in another class altogether.

I agree more nearly with his remarks about the Gresley K2 and K3 mixed traffic types which, he says, are unsuited to heavy coal trains and are apt to be rough riders. The nickname of the latter class, jazzers, bears this out. One must remember that mixed traffic engines are always a compromise. The K3 class, on occasion, work passenger trains very successfully. I once timed 2451, then fresh from the shops, on the afternoon Scotch express from Grantham to York with 500 tons behind on prewar schedule, which was actually bettered. Driver Clough of York was in charge, and as these engines have 5ft 8in driving wheels, this was a really good bit of work.

I admit that on other occasions things have been very different, and I consider the LMS Stanier Black 5s, which have 6ft wheels, two cylinders, and 225lb pressure, better than the three-cylinder K3 engines with 180lb pressure. My correspondent says that the cabs are "always ankle-deep in coal" and have little room for comfortable firing. He criticises the reversing screw and the "bit of string and spanner for the firehole door screen hold back."

I still hold that Sir Nigel Gresley led the way with high-speed, high-power locomotives and I trust that his successor, Mr Thompson, will continue and improve on the Gresley designs.

This led to an exchange of viewpoints with other readers:

Engine types – a further exchange of views

SIR – The article by E.S.Waterhouse (August 1941) in reply to a Peterborough reader says "Stanier 2-cylinder engines are better than K3s," a sure sign of failure of 3-cylinder engines. Being a goods driver with an occasional passenger trip, I do not think three cylinders can touch two cylinders for working. A 3-cylinder will not work suitably or is as capable when notched up as designers imagine. The 3-cylinder makes a bad show working up banks, notched up anywhere above 40 per cent cut off, the 0.4 type easily beating 0.2 type.

Now, as regards Mr Waterhouse on locomotive design (October 1941), does any driver think that the V2 type is comfortable on the driver's side? They may have padded seats, but what about knee room? Also the front window is too far away, and on the slant. This is not good for eyesight; and the side window is a good cause for indigestion. The vacuum brake is too noisy when in "on" position, the small ejector blowing and the regulator continually blowing through. Is that an improvement or a nuisance, especially when coasting with heavy coal trains?

Mr Waterhouse hopes Mr Thompson will improve on the Gresley types and designs. Well, let's hope so.

Hopeful

From ESW's column, December 1941

I was interested in the comments of "Hopeful" in the November 1941 *Journal* on my reference to three and two cylinder engines. Unfortunately, I get little chance of becoming acquainted directly with the work of freight trains, and it may be that they are less suited to that work than to passenger traffic. Mr Stanier's 8000 class, of which engines are now being built for overseas war work, are two-cylindered. Moreover, I find that most drivers prefer the two-cylinder "Riverside" tanks to those numbered 2500-2536, several of which I have noted lately in the St Pancras area, which have three-cylinders. Very few of the three-cylinder type are built exclusively for freight traffic. But the three-cylinder type holds the speed record after all, and has the advantage of less dead centring than the two-cylinder type. On the other hand, it involves more complicated valve gear and costs more to build.

Which led to the following correspondence in the Journal

SIR – Mr Waterhouse's notes on classes of engines state that an LMS Black Five is the ideal mixed traffic engine. It has obviously slipped his memory that the LNER V2 class were primarily a mixed traffic kind of engine, and which LMS Five black-on-red will take up to 455 tons tare and maintain a 72 miles per hour schedule, and notch up inside 15 per cent? This can be done by the LNER's V2s.

No, Mr Waterhouse, think again. For comfort in riding and ease of handling, the East Coast have we "Darbies" in a "flap".

Leeds Liner, LMS

To which ESW replied:

SIR – "Leeds Liner" is right in saying that the LNER V2 class are more powerful than the Stanier Blacks, as they are altogether larger machines with bigger boiler, firebox, heating surface and tractive effort of 33,730 against 25,455. But the weight of the V2 is 148¾ tons against 125¼ tons, and the former are not so useful as mixed traffic engines since they cannot go over as many routes as can the lighter engines.

I cannot understand the reference to the V2 taking 455 tons at a "schedule" of 72 miles an hour; surely a maximum of 72 miles per hour is meant? The 72 mph schedule was that of the pre-war Coronation, drawn by an A4 Pacific with a load about half the above-mentioned. If haulage power is needed the V2 has it, but by the ideal mixed traffic I meant the sort of engine that can go almost anywhere and do almost anything except the heaviest and fastest trains, and the Staniers certainly do it.

E.S.Waterhouse

Letter from the archive

A sort of past-history/real-time account from a Grantham express driver turned King's Cross locomotive inspector, who felt out of place when he moved to York for health reasons.

'Every rise was anticipated, the engine responding like a racehorse that knew its job'

17 Third Avenue
Heworth
York

September 11 1941

Dear Professor

I have felt for some time that I would like to write to you, but have not got any further than that. After reading an article of yours recently I decided to try to get on with it. I have mislaid the journal but if I remember right it was to the effect that a Peterboro driver had suggested that the Ivatt Atlantics were almost as good as the Pacifics.

Some of the engines were known as being very strong, others not so strong but very fast. I was brought up at Grantham, and 3281 was one of the strongest from my early days in 1909. In August 1936 I worked down to York with 3281, assisted by Fireman (now Driver) Harry Johnson of Grantham with 4446. Then, as now, York was very short of power and 4446 was sent to Leeds.

I was given the 2.33pm junior Scotsman, alone. The timing was 89 minutes for 83 miles York to Grantham. The load was 17 cars, 526 tons. The train came in at Platform 14 and the stationmaster was greatly alarmed to see us back on unassisted and said we should never get them started. We did, though, and first time at that. We ran the Normanton line to Chaloners Whin. Having to reduce speed to pass onto the up Doncaster line cost two minutes, and we were two down at Selby.

My fireman was Len Warrington, and he entered into the spirit of it right from the start."Go on", he urged, "let's see what she can do" and we did. Except for Selby Bridge the regulator was never moved from the shop block, full open. The reversing lever was moved up and down according to gradient. Every rise was anticipated and the lever lowered well before reaching it so that speed was maintained uphill. The engine responded magnificently, almost it would seem like a racehorse that knew its job. Both injectors were on whenever on the level or falling gradient, and indeed quite half the journey.

Doncaster had been warned and everyone was ready, but my word we gave them a shock. We regained one minute to Retford, one to Newark, and were on time at Grantham.

That was the best job of work I have ever done and I get a thrill out of it to this day. I think it proves the worth of the engine, but for any man to even suggest that they come anywhere near the Pacifics or V2s is to betray lack of knowledge and experience.

The last time I had a good run as a driver I regained 26 mins from Grantham to King's X with 2557 after brake trouble at Grantham had resulted in failure of 2580, with a very important train, and the last time I had a through run from KX to Newcastle we had 2509 Silver Link, 26 on 740 tons. We lost two minutes to Hatfield, regained it to Peterboro, two to Grantham, refined to York. Had to go easy from Darlington to Newcastle because coal was running short and we cleaned up every ounce of dust at Birtley, but the engine was as good when the last shovelful was put into the firebox as when leaving KX.

I was very happy at KX Loco, had very great responsibilities, and that fact ensured a good standing. Southern Area loco inspectors are Class 2, in the North Eastern Area they are Class 1 and as I was very poorly indeed last year in the London district I applied for a vacancy at York early this year and was successful.

'Here at York there seems to be nothing but chaos'

My health has greatly improved but as far as work is concerned I do not feel quite at home. I have less responsibility and very much less standing. They have no A4s here, and not many in the Area. There aren't many good trains worked from here by York men. The engines are in a very bad state or repair and rarely at their home depot. It is impossible to get necessary equipment – very rarely able to get a sweeping brush to keep coal from under one's feet, and sometimes not even a slacker pipe.

Every engine is in common use practice and there seems to be nothing but chaos and hopeless muddle. Each shed hangs on to any good engine wherever it may be allocated, until it needs something big doing to it, and then passes it on somewhere else. It matters little what may be booked against the engine, it just has to be sent out right away.

Above all, and worst of all, there is very little pride of craft. Nobody bothers, nobody cares. Drivers rarely understand, or at least rarely practice, good engine driving. They have no journal to hand to the guard that can show time lost, so if a bit of time is lost the driver can find some excuse.

As for firemen, well they should find a different name for them here. They just keep on heaping coal into the middle of the firebox and hoping for the best, chimney top full of smoke all day long and, as they rarely bother to slack coal, we all get as black as sweeps.

Many trains have to be assisted from Newcastle to York, then the assistant engine is taken off and Grantham men go forward (and usually keep time) with one. It happens every day. I have just seen the 2.51pm York to King's X off with 20 cars, about 616 tons, assistant engine taken off and Driver Skewith taken on with 2570.

I am doing my best, I talk to the drivers and show the firemen. Many a day I fire almost all the way to show how the engine responds to good firing, for the coal simply must be put along the sides of the box where the water is, if best results are to be obtained, and believe me it is easy. Although I am 49 years of age, thank God I was never fitter.

I conduct all Psd Cleaners 250 hours exams for York, Leeds, Starbeck, Cudworth, Selby, Hull, Bridlington, Scarboro, Pickering and Malton, so have a big district in which to roam about.

'The NE has every engine class ever built, many hopelessly out of date'

One big difference is in the many classes of engines here. The NE seems to have retained every class and type that was ever built and many of them hopelessly out of date. King's X men booked for passenger work had the knowledge that they could only get Pacific, V2 or Atlantic, for they have no B16s or B17s or D49s or K1,2 or 3, to say nothing of the many others we have here. The majority of the D49s are very rough – having no axle box wedges they are awful.

I know I shall feel it when peaceful days come again and I see my late colleague Dick Jones of KX, whom I initiated into the job, going through with the Jubilee and Coronation, but I shall hope to get a look in somewhere.

Should you repeat or use any of the foregoing I know you will not disclose the source of the information, for though it is an understatement it might not help me if my employer knew I had broadcast it.

By the way, streamlining is being taken off in the shops and not replaced at present. Have you been on the Bantam Cock yet? It is a grand bit of work.

Yours sincerely

G.Manyard, Loco Inspector
Chief Running Supts Dept, York

Further correspondence shows that Manyard left York to return to King's Cross as District Loco Inspector.

Giant new postwar engines unlikely

October 1941

It is interesting to speculate about the postwar types of locomotive that may appear on British lines. In the United States the 4-6-4 type fitted with a booster is popular on most lines, whilst 4-8-2 and 4-8-4 express engines are not uncommon. Our short turntables have always been a hindrance to larger engines being built, though the obstacle is not insuperable. It is our cramped loading-gauge which gives little hope of really giant locomotives in this country.

In France there are – or were – some magnificent compounds with four cylinders and variable cut-offs for both high and low-pressure cylinders, which enabled drivers who had the skill to get wonderful results. In this country, largely because of the loading-gauge which limits large outside cylinders, the compound seems to have no future.

Sir Nigel Gresley, however, did give us the powerful 2-8-2 engines and, by turning the experimental engine (WI 10 000) into a 4-6-4 simple, provided two designs which may be more prominent on other lines in the future.

If it were possible to build a boiler big enough to supply sufficient steam, the 4-6-4 with a booster would be an ideal engine for heavy express work. The booster enables a good start to be made and can be cut off automatically at 30 miles per hour. True, the experiment with Atlantic 4419, which once had a booster, was a failure, but the boiler was too small to warrant it. The booster fitted to the 2-8-2 P1 freight type was also, I believe, removed.

But in America the acceleration given by the booster must be seen to be believed and sharp point-to-point timings such as existed on the Great Central section from London northwards are made possible much more easily by such a fitting. The bogie under the cab makes the 4-6-4 a much more comfortable riding engine than the 4-6-0, and the success of 10 000 – the only one of this type in the country – deserves repetition.

Whether we shall see pulverised coal and automatic firing is a more open question but, if we could do more in this country to extract oil from coal, oil fuel would be a distinct possibility. My readers who suffered during the coal strike with engines improvised to burn oil must be reminded that I am speaking of fireboxes designed for oil burning such as Mr Holden fitted in bygone days to some Great Eastern engines with success.

Letter from the archive

LONDON REGIONAL COMMITTEE FOR EDUCATION
AMONG HIS MAJESTY'S FORCES

At London School of Hygiene
and Tropical Medicine,
Keppel Street,
W.C.1.

Museum 3041

29th September, 1942.

Dear Professor Waterhouse,

I was talking to Mr. Hind of the British Railways' Press Office and I happened to mention to him that railway engine footplate riding was a hobby of yours. He writes that he would be glad to give you any assistance in the way of collecting material, either factual or photographic, if you consider talking on your railway experiences. If you are interested, I suggest you get in touch with him direct. His name and address are:-

H.R. Hind, Esq.,
British Railways' Press Office,
c/o South Railway,
Waterloo Station, S.E.1.
Tel: Whitehall 6146, Ext. 64.

Yours sincerely,

Harry Ross

Professor Waterhouse,
Principal's House,
Richmond College,
Surrey.

HR/CJH.

There is no confirmation ESW took up this offer, perhaps he didn't care for the word 'hobby'

Signals, schedules and obstructive freight

November 1941

A recent trip to Weston-super-Mare was notable chiefly for the abundance of signal checks, seven each way between London and Bristol. We are said to be short of coal, but no one would think of it, judging by the long coal trains I saw on that trip. At Didcot four were drawn up nose-to-tail west of the junction.

As we passed Swindon I saw the old Brighton Terrier 2643, No 5 of the Weston, Cleveland and Portishead line. Originally No 43, Gipsy Hill, she was built in 1877, and I fear that the old lady, after 64 years of service, is now on the scrapheap.

To Bristol I travelled in a 403-ton train behind King class engine 6004, and Messrs Hiscox and Middleton of Old Oak shed managed to keep pretty close to schedule all the way despite the obstructions of the signalmen.

On the return early next day I found two old friends in Messrs Amos and Treago of Weston shed with a Manor class engine working the train as far as Bristol, and I was pleased that though it is eight years since we last met I was recognised at once. A Castle class, 5022 with 365 tons, was smartly worked up from Bristol by Messrs Hewitt and Kimber of Bath Road shed; the last 36 miles from Reading – passed very slowly because of signals outside – was done in 45 minutes, including a dead stop at Slough and further signals soon after, and Paddington was reached just to time.

On another run to Birmingham with a similar engine and load, 5031 and 357 tons, the hindrance was pw checks, four of them; but Driver A C Hunter, with whom I have travelled before, and Fireman Southan, of Wolverhampton, kept moving well, and but for a stop outside Birmingham would have delivered me there right at time. The journey was made pleasant by the company of one of our members, Fireman Askett of Chichester, a grandson of one of the old Brighton drivers who drove Atlantic No 39 in the days when each man had "his" engine.

On the return trip I was in with a jolly crowd of six of our Birmingham members as far as Leamington, and we talked shop all the way. This train is very easily timed, the non-stop run up from Leamington calling for an average 45 miles an hour. Driver F Taylor and Fireman Smith, of Wolverhampton, with a good King (not all GW Kings, as well as other kings, are good) and 391 tons had an easy time. But I admired the way in which the trip was run, never too much ahead of schedule and yet always a few minutes in hand to provide for checks, which worked out at three pw and one signal check, and we arrived a leisurely three minutes ahead of time.

Another run took me to Mansfield. From Marylebone Driver F S Smith and Fireman Worthington, of Neasden, had easy work with a nine-coach train and Pacific 4474. They had time in hand all the way to Leicester, where a friend I have not seen for a long while – Driver Webb of Leicester – and Fireman Bennett took over. The load was now 12, and despite a two-minute-late start and a pw check, Nottingham was reached to time. Another member – Driver Eaton of Annesley – took me on with an ex-GC Atlantic.

I returned by LMS, and was interested to find an ex-LTS 4-4-2 tank working on the local train to Nottingham and taking 200 tons on the up and down road with a number of stops easily in the booked time. The journey on was behind Stanier three-cylinder 5641 with 11 bogies, and was capitally done by Driver Askew and Fireman Dove of Nottingham.

At one point we were four minutes behind, and time was overspent at almost every station; but the running was so sharp that everything was regained, one particularly good sprint being the seven miles start to stop from Kettering to Wellingborough in just nine minutes. We passed St Albans with ample time in hand, but a pw at Mill Hill, followed by a long crawl at 15 mph for relaying in the Haverstock Hill tunnel, more than accounted for the two minutes out with which we finished.

It is plain from these runs that, given anything like a clear road, war timings are liberal enough to ensure exact punctuality on most trains; but on some lines especially, the problem of sandwiching in the freight trains has not been solved – and in that case the most liberal schedules do not help much.

EDITOR'S FOOTNOTE

After the panic of the Blitz, reset railway schedules were realistic. Enginemen settled to the job in hand. ESW makes a special point of mentioning old acquaintances.

Driving Flying Fox when Sir Nigel Gresley chose to join me on the footplate

47 Hitchings Way
Woodhatch
Reigate
Surrey

July 27 1942

Dear Sir

I hope you won't be bored with a few incidents which happened during my career as a driver, mostly on the Pullman. My old friend Sparshott and I used to kind of compete with each other to try to get the best run. Whatever the delay at the starting point or on the trip I always did my best to arrive at the destination on time if possible (Sparshott did the same).

I had a wide and varied experience with the different classes of engine. Atlantic 4450 was, I think, the most economical engine I had on the Pullman. I had 4419 for some time with the booster fitted. I believe it was the only express engine in the country fitted with the booster but unfortunately I experienced great difficulty with the steam joints blowing out about twice a week, causing the cab to be filled with steam for the rest of the trip.

I then had the four cylinder Great Central engine Earl Haig, master of anything but very heavy on coal. I had the honour of taking Her Majesty Queen Mary to Harrogate in 1926 with the above engine.

After being transferred to the Flying Scotsman I was allocated Flying Fox, which was capable of excellent work. I remember on one occasion, after passing Peterborough on the Flying Scotsman, Sir Nigel Gresley walked through the corridor onto the footplate and was agreeably surprised at how I was working the engine. He wanted to know if she would keep time in the position she was. I asked him to make use of my seat and fancy he was the driver of the train. He stayed on the footplate until we took water at Scrooby. I felt sorry when he left me as we had the most interesting talk on engines that I ever had (wonderful engines those Pacifics).

Hoping you are keeping well
I remain
Yours truly
Wm Barnes

Retired without an enemy, never late on duty

53 Ponsford Road
Knowle
Bristol 4

December 8 1943

Dear Sir

Just a few lines to let you know how surprised and pleased I was to see my name in the *Journal* for this month, especially since I retired at the end of September. Also of my mate G Peacock with my favourite engine (5025).

Well, Sir, I think I can safely say that when I left the GWR I left without an enemy, not only in my own grade but in all grades I came in contact with during my course of duty. I did 43½ for the company, during which time I am proud to say I never had a bad fireman and was never late on duty.

Yours truly

W.Y.Bright

PS I think I ought to tell you that my reason for retiring was that if I had stuck to my post I should be keeping the likes of my own mate a lot longer on the shovel. So Cheers, Best of Luck.

SOUTHERN RAILWAY.

TELEPHONE:
Dorking 3201 Ext.9

LOCOMOTIVE RUNNING SUPERINTENDENT'S OFFICE,
DEEPDENE HOTEL,
DORKING, SURREY.

IN YOUR REPLY PLEASE QUOTE THIS REFERENCE:
HO.67160

YOUR REF.:—

22nd December 1943

Rev.Professor E.S.Waterhouse,
The Principal's House.
Richmond College.
RICHMOND.Surrey

Dear Professor Waterhouse,

Yours of December 18th: I must apologise for failing to return the second copy of your draft, which was filed on my papers by mistake. I return it herewith, together with one or two pencilled amendments.

I mentioned, in connection with the first draft for the A.S.L.E.F.Journal, that steam versus screw reversers were a matter of opinion because there are quite a number of men who prefer the wheel-and-screw on express engines. There is not a great deal of difference as regards manual effort, especially as these engines are not used for shunting. The wheel is positive and, generally speaking, more reliable than the steam reverser which may not hold in position if there is a slight leak in the oil locking cylinder and may be very slow in action in certain circumstances. On an engine which is used to any extent for shunting a power-operated reverser is a boon, but, as I mentioned before, on express engines they form a bone of contention and I therefore felt/advisable to point this out to you. As you know, we have a wheel reverser on the Lord Nelson, Schools and King Arthur engines which meets with general approval and, provided it is properly oiled, gives no trouble or undue labour in operation. /it

Of course, the longer one lives the more one realises the truth of the saying that "One man's meat is another man's poison".

Wishing you the compliments of the Season,

yours faithfully,

Encl.

This letter from a Southern Railway official, offering amendments to ESW's draft article on a footplate trip, discusses issues surrounding steam versus screw reversers used for cut-off control of boiler pressure from the footplate. It shows how seriously railway companies took ESW's assessments

HON. SECRETARY:
H. PURSHOUSE.
11, WHITE LEE GROVE.
MEXBOROUGH.

MEXBOROUGH LOCO MUTUAL IMPROVEMENT CLASS

E.S.Waterhouse. Esq.,
c/o The Editor.
The Locomotive Journal.
9, Arkwright Road.
Hampstead. N.W.3.

16, Beechville Avenue.
Piccadilly.
SWINTON.
Mexborough.
17th March 1944.

Dear Sir,

Having read your articles in the "Locomotive Journal" over a number of years, I would like you to know how I have enjoyed reading the same, at the same time I would like to register a mild protest.

Every article I have read up to the present has dealt exclusively with engines and crews of express and ordinary passenger trains, with men who are always in the public eye, please do not mis-under stand this for a bit of professional jealousy, far from it, but these men have got the best of everthing to work with, /such as it is/

What about the men who work the humble coal and heavy freight trains under the worst possible conditions, long hours, bad coal, black-out, leaking engines, etc, etc. Are they not worthy of being mentioned in despatches. The passenger sees them, if only from the station platform.

I respectfully suggest you take a trip on one of the jobs at my own depot, i,e. banking heavy coal trains from the Wath L.N.E.R. Concentration Yard with an ex G.N.Rly.0-6-2. suburban tank engine to Dunford Bridge, these engines are called, /among other things/ "Dive Bombers" and the mileage of these banking jobs are a mere 21 miles but often seems like 100.

Believe me Sir, you would get sufficient experiences to fill a volume of the "Locomotive Journal".

In conclusion, let me state, whether you mention the humble side of our profession or not, I shall continue to read and enjoy your writings.

Wishing you continued success.

Yours faithfully,

Driver. Mexborough Loco.

This letter respectfully reminds ESW that express journeys are far from the only service enginemen provide for the nation. He wrote rarely about the freight side because his usual involvement was as a passenger. The letter shows how keenly drivers took their craft. Mutual improvement classes were found throughout the rail system

This correspondent wrote anonymously, obeying the instruction that serving NCOs should not disclose their whereabouts

Anonymous news from the LNER's North East

May 15 1944

Dear Sir

Humbly I beg your pardon for writing in such a manner, but I perceive in the *Locomotive Journal* so kindly sent by friends at my home depot that it is some time since you visited the North Eastern area of the LNER. Therefore I hope you will not take offence at this little example of the work being done in this area.

On April 12 it was my pleasure to travel to York by the 10.25pm Relief from Newcastle-upon-Tyne. Strolling along to the engine I found it to be an A4 4466, Herring Gull, with Bro Todd and Bro Keay of Heaton in charge. Load 18 coaches for 579 tons.

A late start of two minutes was regained by the time Durham was reached, then a distant signal checked our progress up the stiff gradient out of Durham, resulting in four minutes being lost. Darlington was reached a minute late and was left almost five minutes late, but by the skill of our two Heaton members York was reached only one minute late.

On the return journey with the 2.18am from York the same crew manned engine A4 4468, Mallard, the load being 16 "bogies" as the guard called them, 509 tons all told.

We left York eight minutes behind the clock but 4468 is famous for her running powers and we were past Darlington in 44 minutes. This meant we were only four minutes behind, and Mallard was going like the wind. Ferryhill was reached in 16 minutes from Darlington, this gained another two minutes, but at Tursdale we were brought to a dead stop by signals.

Then we passed Durham at the slowest speed possible for some reason unknown to me, costing six minutes in all. However, from then it was like the prewar running of trains. Brothers Todd and Keay did their utmost to gain a little of those six minutes, and succeeded in regaining four minutes.

On asking Bro Keay what he thought of 4468 he said "The best I have ever been on!" I say "Well done, youngsters", for Bro Todd is only 41 and Bro Keay 22.

Sgt Maj ASLEF, Royal Engineers. Ex LMS Camden Town
Somewhere in Scotland

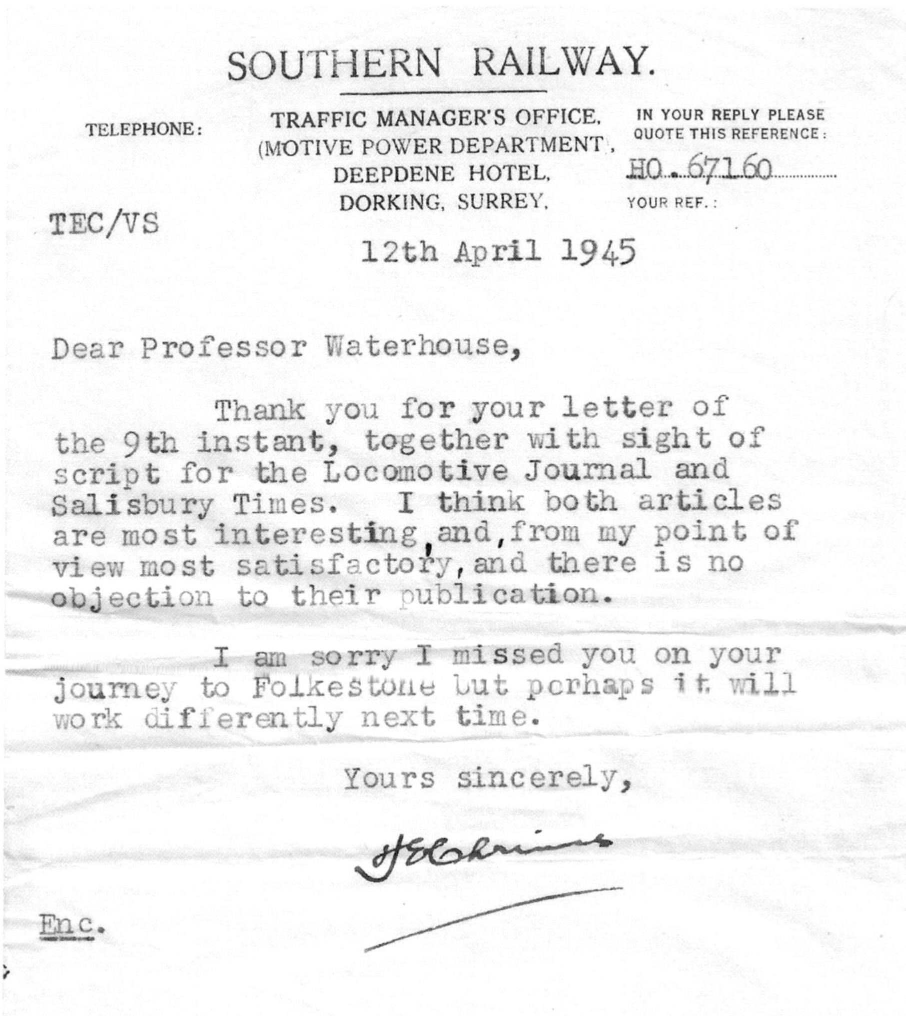

SOUTHERN RAILWAY.

TELEPHONE:

TRAFFIC MANAGER'S OFFICE.
(MOTIVE POWER DEPARTMENT),
DEEPDENE HOTEL,
DORKING, SURREY.

IN YOUR REPLY PLEASE QUOTE THIS REFERENCE: HO. 67160

YOUR REF.:

TEC/VS

12th April 1945

Dear Professor Waterhouse,

Thank you for your letter of the 9th instant, together with sight of script for the Locomotive Journal and Salisbury Times. I think both articles are most interesting and, from my point of view most satisfactory, and there is no objection to their publication.

I am sorry I missed you on your journey to Folkestone but perhaps it will work differently next time.

Yours sincerely,

T.E.Chrimes

Enc.

This letter from T.E.Chrimes, Southern Railway's Superintendent of Motive Power, gives his blessing to the May 1945 article overleaf. Interestingly, Chrimes also okays a piece for the *Salisbury Times* which ESW evidently produced from the same material. After all, it was the result of five Salisbury round trips

Five Salisbury returns in late wartime

May 1945

Here are some notes on five return trips to Salisbury on the 2.50pm down and the 9.50am up. Distances: London to Basingstoke, 48; Andover 66½; Salisbury 83¾ miles. The time in each case is 112 minutes, but the up train does not stop at Basingstoke. Load, except where stated otherwise, 13 bogies, about 410 tons.

Down trips worked by Salisbury men. Up trips by Nine Elms. Engines – down: (1) Pacific 2.1.C8, Bros Hopkins and Sidgwick; (2) 2.1 C6, Bros Lawford and Gooderidge; (3) 2.1 C7, Bros Johnson and R.Steele; (4) Arthur (N.15) 4-6-0 455, Bros Head and Kightley; (5) 2.1.C10, Bros Hopkins and Bartlett. Up trips: (1) 2.1.C7, Bros Coward and Hollis; (2) N.15.791, Bros Mills and King; (3) N.15X (former LSBC tank rebuilt as tender engine) 2333, Bros Berridge and Aplin; (4) N.15.456, Bros C Smith and Beresford; (5) 2.1.C7, Bros G Lark and Meech.

Down trips: (1) Much pw and signal delays, so we took 23 minutes for the first 12 miles, and five minutes overtime at stations, added to a pw at Byfleet, together with poor coal, made arrival nine minutes late, actually a net gain of three minutes; (2) 15 bogies. Delays at stations and by pw 15 minutes, of which seven were regained by excellent work; (3) 16 bogies. Delays amounted to 14 minutes, of which three were regained – good work with this load; (4) the old and smaller engine did valiantly, regaining three out of ten minutes' delay with 412 tons; (5) by the courtesy of Mr T.E.Chrimes, Superintendent of Motive Power, I was permitted to ride on the footplate. We had an engine fresh from the shops. After Raynes Park, cut-off was fixed at 15 per cent all the way to Basingstoke. Despite four minutes' loss by signals and pw and a dead stop outside Basingstoke, we arrived at 3.49 – two minutes early. The up-and-down run to Andover was done easily in 21½ minutes. From there to Salisbury in the same time, arriving nearly two minutes ahead of schedule. The engine was at 15 per cent all the way except at starting, with half regulator or less, and played with the load. On every one of these trips time was recovered.

On the up trips the Pacifics showed to advantage against the smaller engines much more markedly: (1) Start was ten minutes late; two minutes were regained to Andover and lost by station time, but from Andover the run was made in just under 75 minutes, including a signal check almost to a dead stop at Wimbledon, and we arrived one minute early. It was plain that much more time could have been gained had it been needed. As it was, schedule was bettered by at least 15 minutes; (2) Start and arrival seven minutes late, with signal and station delays amounting to six minutes recovered; (3) a smaller engine with 180lb pressure, as opposed to 200lb on the N.15s and 280lb on the Pacifics. Wheels 6ft 9in as against 6ft 7in and 6ft 2in, less weight and tractive effort. Nonetheless, the schedule was kept, and indeed bettered, by about three minutes; (4) this was a clear run, one minute regained; (5) I returned on the footplate. Nine Elms has just got any of this class stationed there. Driver Bro Lark has handled them only twice before and Fireman Bro Meech was having his first experience on this sort of engine. Indeed, Fireman Bro Kiddle, of Salisbury, fired to Andover to show him the ways of this type. So good was the instructor, and so apt the pupil, that on his taking over things went excellently.

All previous trips started about ten minutes late. Here we were but one minute behind, reached Andover to time and left one minute late. Although cut-off was 15 per cent, with easy regulator, the engine ran ahead like a racehorse, just as it had done on the first trip. Then we had a late start. Now we were getting too early. Steam was shut off on the down grades to ease the engine, but short of applying the brakes nothing seemed to check her.

A train ahead of time is almost certain to be checked by signals, but that simply would not happen except for one distant at Esher, which went off at once. We had the green lights all the way up and finally ran in 11 minutes before time – surely a wartime record!

EDITOR'S FOOTNOTE

A classic example of ESW's meticulous timekeeping, along with other notes, enlivened by two footplate trips. Proof that the Southern's Waterloo-Salisbury service remained in relatively good shape as the war drew to a close.

1946-1964

'My own opinion, which most LMS men will share, is that for fast passenger, stopping and freight work, the best type in the country is the Stanier Black 5000 class'

April 1946

'Good and bad engines seem to retain their character from beginning to end of their career. Sometimes a good one loses its virtues, but it is rarely that a poor one shows much improvement'

July 1956

'To close lines without proper and equally convenient other transport is to treat railways like multiple stores – pay or close'

November 1963

'Does the average passenger greatly care if, let us say, a 300-mile journey takes three or three-and-a-quarter hours? Is not a comfortable journey, punctual, at reasonable cost, more likely to attract him than very high speed?'

January 1964

'We are hoping that in our modest fashion we will assist to revive that pride in our craft which is receiving such very hard knocks'

February 28 1946

The Locomotive Express
3 Eltringham Grove
Edinburgh 11

Dear Sir

I was greatly pleased to see your letter regarding our modest publication. I have read your articles with a great deal of interest each month in the *Journal* over a number of years, and will be only too pleased to place your name on the mailing list.

I know you will be interested in the *Express*, as we consider E.S.Waterhouse if not one of ourselves at least very near to that. Your contributions in the Journal prove that you enter into the very spirit of all our problems with sympathetic understanding.

I feel that you would thoroughly enjoy the atmosphere in the few square inches of office space which constitutes the editorial sanctum of the *Express*, which is being run by locomotive men with that enthusiasm which seems to be our mainstay. At all hours of the day and night some of us are here, joined by an occasional policeman, and out of the arguments are bred much of what appears with the pages of the *Express*.

We started off with the most modest ambitions, but are finding that paper restrictions are the real stumbling block to our circulation, as with every post requests come in for more copies from every part of Britain. We are hoping that in our modest fashion we will assist to revive that pride in our craft which is receiving such very hard knocks from every quarter at the moment.

May I express my very sincere pleasure at personally contacting you – I have wanted to do so for years, but couldn't pluck up the courage.

Yours sincerely
Norman McKillop (Toram Beg)

'It seems absurd to have the whole nation held up for a few feet of snow'

March 11 1947

Dear Sir

Many thanks for your very kind letter. Nothing could give me greater pleasure than to meet you on your visit to Edinburgh. I'm looking forward to the month of May.

Thanks also for the suggested contribution to the *Express*. It will be like meeting an old friend in a new medium to most of our readers if you can find time to put the suggestion into effect.

The weather in Scotland is still very bitter. Am on the Edinburgh-Perth trip this week. Very cold with gusty snow and sleet most of the last two days, but it is a let-up from the long hours on the East Coast trains. Got home on Sunday morning without having been to bed for over 24 hours. Simply tuckered out. Turned in, accompanied by two hot water bottles, and lay like a dog for 22 hours' solid sleep. I cannot remember feeling so done up after a week's work, but my case wasn't the worst by long chalks – my mate on the opposite turn was snow-bound for 20-odd hours and is still feeling the effects.

I am beginning to feel a little angry about the whole business. It seems absurd to have the whole nation held up for a few feet of snow – when one considers the extremes which are met in other countries where proper measures are taken for dealing with the conditions.

Best wishes
Yours sincerely
Norman McKillop

EDITOR'S FOOTNOTE

Norman McKillop contributed to the Journal *under the pen name of Toram Beg. He was an Edinburgh-based mainline driver who started his own journal,* The Locomotive Express, *and was chosen to write the story of ASLEF's first 70 years, published as* The Lighted Flame *in 1950. His friendship with ESW is recorded here and on the following page*

'Nine-tenths of modern firemen personnel detest the very idea of shovelling coal'

April 5 1948

Dear Mr Waterhouse

Many thanks for your letter of the 2nd, and I was greatly interested in your remarks on the Diesel. As you say, I do get a tremendous kick out of handling the Gresley Pacific, but I musn't forget that the railways aren't staffed by Waterhouses or Toram Begs. Possibly you haven't as close an opportunity as I have of sensing the modern attitude to anything in the nature of labour – nine-tenths of modern firemen personnel detest the very idea of shovelling coal.

Let me tell you a yarn I heard in Dundee only yesterday. He was a real old "shell-back" from Aberdeen, deeply rooted in his job as a driver. He was being hauled over the coals by his Shed Master for the loss of time through steam shortage:

"Was your fireman a competent man?" asked the SM.

"Firemen", snorted the old chap, "there's nae compeetant fireman noo-adays...
"I've had them wi' gold rings on their fingers, and broon boots, and gloves, maybe if you could yin wae earrings as well that micht mak' him a fireman."

Perhaps the old fellow was prejudiced unduly against the modern lad, but it certainly gives one a line on the present situation, and I'm not going to say that I haven't a sneaking sympathy for a desire on their part to cut out the hard work. If, or when, you write a short screed for the *Express* – why not let yourself go on the theme. Possibly half the locomotivemen in Britain will come down on you, but there's nothing like having an opinion and stating it. That's why the *Express* was brought to birth.

All best wishes
Sincerely yours
Norman McKillop

PS (handwritten) You may be interested to know that I've been requested to write up the history of the A.S.L.E.& F. The EC passed a resolution to this effect a few months ago and I'm coming to London occasionally to do some research for this purpose. NMcK

Stanier Black 5 the best all-rounder

April 1946

My own opinion, which most LMS men will share, is that for fast passenger, stopping and freight work, the best type in the country is the Stanier Black 5000 class, with a 6ft wheel, especially the latest, numbered 4800-4900. They have timed up to 92 mph, and they'll make a good showing at almost any job. They cannot be cut off as soon as most express engines, but at 20-25 per cent they work well and economically.

Take this trip, for example: Engine 4828, load 13 about 380 tare, Bros Twyford and Garrod of Kentish Town in charge. On the 4.05pm from St Pancras the first stop is Kettering (72 miles) in 91 minutes. On leaving we were stopped twice for electric signal failure at Brent, pw at Elstree, dead stop at Chiltern Green, so that the first 30¼ miles, to Luton, took over 50 minutes; at Oakley, a pw for bridge repairs, at Wellingborough signals, and a stop outside the station whilst a freight train leisurely put over. Yet the total loss was but 12 minutes. By Nottingham, another pw and signal stop for two minutes and drawing up at every station, twice indeed at one, made us 23 minutes late, but with three coaches less that was reduced by five minutes to Chesterfield, where I got off. It is evident that lively running was the order of the day whenever we were given the road.

Returning from Nottingham I had evidence what difference a good engine makes; 5277 (Bros Mitchell and Low, Kentish Town) had the same load, but as signs of trouble were evident, took on a 4-4-0 (Bros Armstrong and Wilcox of Nottingham) as a "hooker", and all the way up it was a struggle to keep to schedule – a bit lost here and got back there – despite the double heading. I heard later that the same engine had played up another driver on the Scotch train a few days before. Not even the best of engines will stand up indefinitely to modern neglect, and it is amazing what they do stand.

Speaking of the mixed traffic type, I admit the good qualities of the GWR Halls, though my first vote goes to the Staniers. The best work with very heavy loads during the war was that of the LNER Pacifics, and what was most strange was that some of it was done with the unrebuilt A1 class of 1922, with only 180lb pressure. The best 4-4-0 engines are the Southern Schools, although I still have respect for an LMS compound – especially in the hands of Scottish enginemen. Perhaps I will have to revise these estimates. But as war conditions indicated them, these were the best.

An old Director delivers the goods

May 1946

On a journey to Sheffield I went from King's Cross with V2 3662 (Bros Anthony and Gibson of Peterborough) with 14 on, running right to schedule. From there a similar engine, 4899 (Bros Footit and Gutteridge of Doncaster) was not climbing well, and dropped four minutes to Grantham, but two were regained by smart running to Newark; pw delays spoilt the running to Retford. From there I went on with two Grimsby members, Bros Whittaker and Heaton, with Sandringham class 2847, which was well able to do what was needed on a stopping train to Sheffield.

I returned on the new breakfast car express to Marylebone. One of the old Director D10 class, 5430, was hauling eight bogies, about 260 tons. Bros Malley and Hall of Sheffield made light of the easily-timed run to Nottingham and, though stopped outside, were in five minutes early. Bros Beck and Hutchinson of Leicester worked on to that city to time. Then Bros Healey and Clifford of Leicester took over. Between Rickmansworth and town we had three dead stops but arrival was only three minutes late. The old engine steamed well all the way, and was easily on top of its work. It is quite a long while since I was behind one of these engines, which did fine work 25 years ago and are evidently still able to deliver the goods.

Travelling to Stoke on a relief train of 14 bogies, Stanier Black 5 4938 (Bros R.Evans and Ball of Longsight) left Euston 25 minutes late and regained eight minutes by smart work to Rugby, where we were held up for five minutes by signals. Getting away, we passed Lichfield – 34 miles in 38 minutes – but after that signals and stops ruined what might have been a great run.

Returning with a similar engine, also a relief train of 398 tons, Bros J.Gould and Nicholson of Longsight did well with 5352 and despite two dead stops bettered schedule of 146 miles in 187 minutes by two minutes, net time being about 160 minutes. It is evident that the faster schedules coming in this month can be tackled, but unless a clearer road can be obtained than one gets on one's journeys now, there will be no semblance of punctuality.

By courtesy of Mr Mason I went over Stoke sheds during my visit and was interested in the 111 engines shedded there. Crab tanks are prominent, together with standard 4 class goods, one or two Stanier Blacks, and some of the old "cauliflowers" of the LNW which haul ballast trains.

Letter from the archive

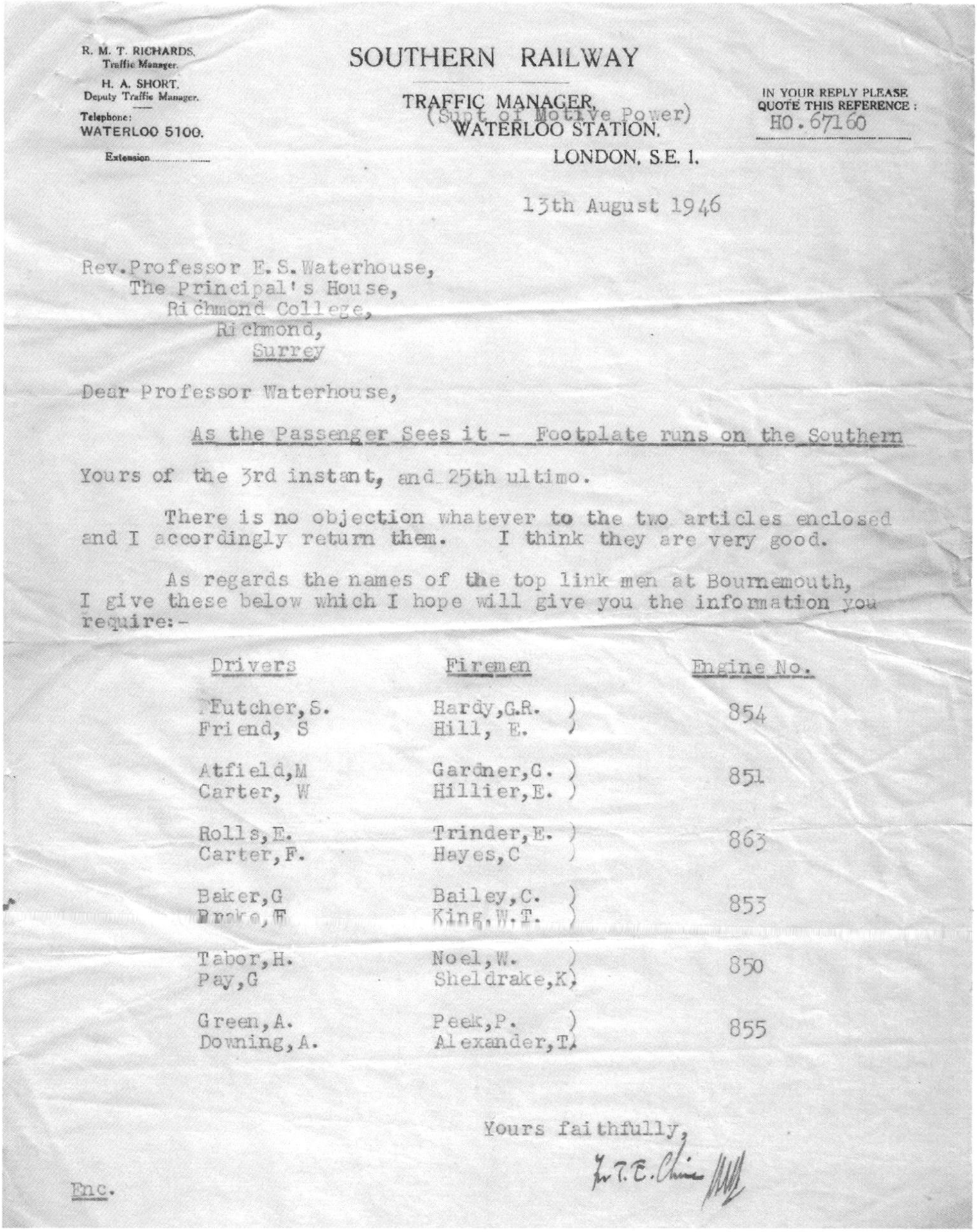

R. M. T. RICHARDS,
Traffic Manager.
H. A. SHORT,
Deputy Traffic Manager.
Telephone:
WATERLOO 5100.
Extension..............

SOUTHERN RAILWAY

TRAFFIC MANAGER,
(Supt. of Motive Power)
WATERLOO STATION,
LONDON, S.E. 1.

IN YOUR REPLY PLEASE QUOTE THIS REFERENCE:
HO. 67160

13th August 1946

Rev. Professor E. S. Waterhouse,
The Principal's House,
Richmond College,
Richmond,
Surrey

Dear Professor Waterhouse,

As the Passenger Sees it - Footplate runs on the Southern

Yours of the 3rd instant, and 25th ultimo.

There is no objection whatever to the two articles enclosed and I accordingly return them. I think they are very good.

As regards the names of the top link men at Bournemouth, I give these below which I hope will give you the information you require:-

Drivers	Firemen	Engine No.
Futcher, S. Friend, S	Hardy, G.R. Hill, E.	854
Atfield, M Carter, W	Gardner, G. Hillier, E.	851
Rolls, E. Carter, F.	Trinder, E. Hayes, C	863
Baker, G [illegible]	Bailey, C. King, W.T.	853
Tabor, H. Pay, G	Noel, W. Sheldrake, K	850
Green, A. Downing, A.	Peek, P. Alexander, T.	855

Yours faithfully,

[illegible]

Enc.

This letter refers to the article on the following spread about the Bournemouth run. Confusingly, the names of top Bournemouth link drivers and firemen supplied do not tally with those in the article, who were based at the Nine Elms shed in London

Merchant Navy v Lord Nelson to Bournemouth

September 1946

Most of my runs are observed from the coaches, but the passenger cannot "see it" properly from that angle and, to appreciate better the work about which I write, I obtain permission at times to view it from the footplate. Mr T.E.Chrimes, Superintendent of Locomotive Power, Southern Railway, has always been most courteous at enabling me to write for members from this vantage point. By his permission I went to Bournemouth and back and was able to compare the running of the Merchant Navy and Lord Nelson classes which now do most of the Bournemouth express work.

On the 10.30am from Waterloo, Merchant Navy 21C12 was handled by Driver Till, who has been heard more than once on the wireless, and Fireman Budd of Nine Elms. A load of 361 tons was child's play, and as we were getting too much ahead at Basingstoke the engine was eased and mounted to Worting summit on a mere wisp of steam. Normal cut-off throughout was about 18 per cent, and practically never more than ¼ to ½ regulator. Southampton (79 miles) was done in 90 minutes – 97 being booked – but we wanted more time to take up water there. The 29 miles on to Bournemouth was reached in 35 minutes, arrival being five minutes early. Running time was therefore only 125 minutes with the stop and so much in reserve that given a reasonable load and a non-stop run these engines could maintain a 105-minute schedule without difficulty.

On the return, Driver Cummins and Fireman Boniface of Nine Elms had Lord Nelson class 857. These are the 4-cylinder, 8-impulse type of 4-6-0, and partly on account of the impulses it was a smooth-running engine. Load was 429 tons, with one stop at Southampton. No difficulty was experienced in timekeeping, though in the latter part of the run the fire gave a little trouble and needed the use of the irons. The long climb of 1 in 250 from Eastfield to Lichfield tunnel – some 20 miles – was done with 20 per cent cut-off and ½ regulator unchanged. After the summit was reached, cut-off for the rest of the journey was 15 per cent. Waterloo was reached two minutes early. Both engines were freshly-painted green, clean, and in "good nick", so everything suggested peacetime conditions again.

I am more and more impressed by the Merchant Navy class. They have both speed and power. The design of the valve gear seems to give full compensation for the smallish driving wheels (6ft 2in), and the engines are fast, both up and down the banks. The steel fireboxes seem quite satisfactory. The electric lighting is a great boon in seeing gauges. The steam-operated treadle-worked fire doors enable the fireman's job to be much more easily done.

These engines were a bold challenge to conventional design in many ways, but they certainly have made good, and for express passenger or freight work and general adaptability they are unsurpassed by anything in the country at present.

Letter from the archive

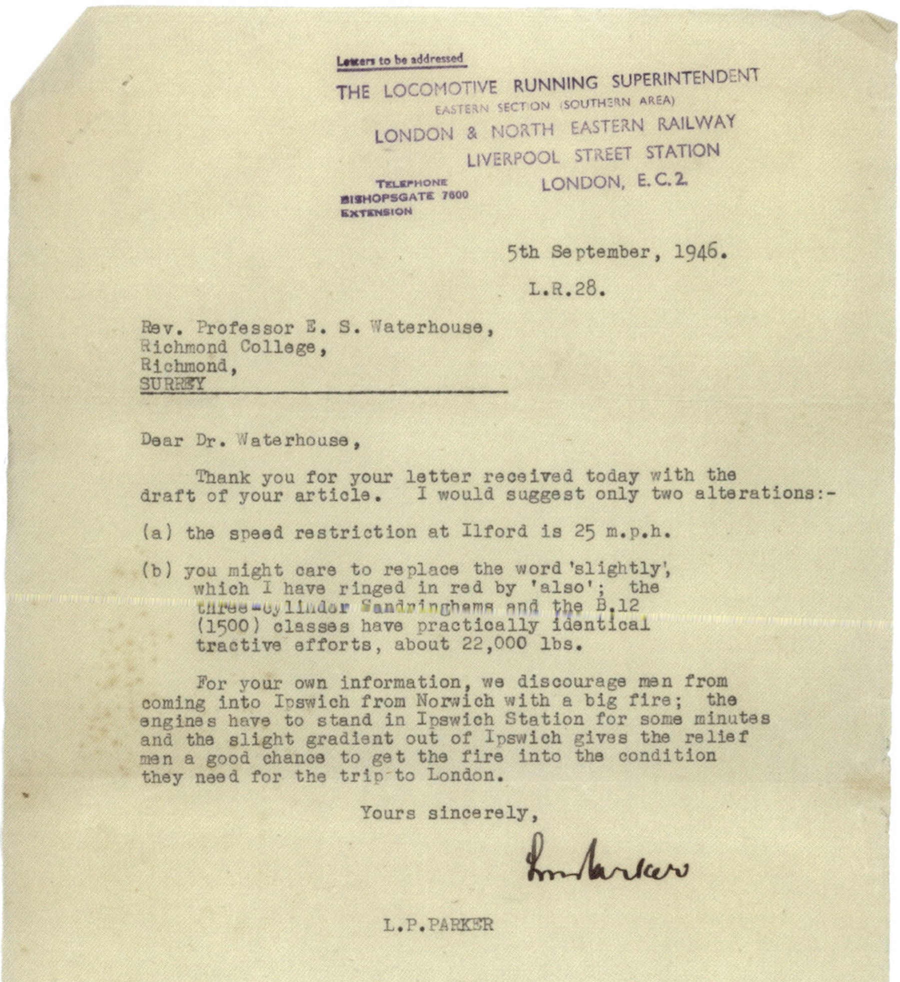

Letters to be addressed

THE LOCOMOTIVE RUNNING SUPERINTENDENT
EASTERN SECTION (SOUTHERN AREA)
LONDON & NORTH EASTERN RAILWAY
LIVERPOOL STREET STATION
LONDON, E.C.2.

TELEPHONE
BISHOPSGATE 7600
EXTENSION

5th September, 1946.

L.R.28.

Rev. Professor E. S. Waterhouse,
Richmond College,
Richmond,
SURREY

Dear Dr. Waterhouse,

Thank you for your letter received today with the draft of your article. I would suggest only two alterations:-

(a) the speed restriction at Ilford is 25 m.p.h.

(b) you might care to replace the word 'slightly', which I have ringed in red by 'also'; the three-cylinder Sandringhams and the B.12 (1500) classes have practically identical tractive efforts, about 22,000 lbs.

For your own information, we discourage men from coming into Ipswich from Norwich with a big fire; the engines have to stand in Ipswich Station for some minutes and the slight gradient out of Ipswich gives the relief men a good chance to get the fire into the condition they need for the trip to London.

Yours sincerely,

L.P.PARKER

Another permissions letter, from a LNER official making specific points about ESW's submitted article, followed by an example of the company's running advice to enginemen

Postwar schedules tougher than prewar

November 1946

For punctuality it is essential that the schedules allow a margin for the recovery of the delays met on almost every run. But my recent experiences show that, generally speaking, neither engines nor coal are good enough to do more than barely keep present schedules and, as a lot of pw work is going on, almost all journeys end late.

I had four trips on the LNER with the same load of 15, say about 470 tons. Pacific 29 climbed painfully slowly from King's Cross, losing time to Hatfield, but from passing that point kept time to Peterborough and lost again on the climb to Stoke summit. Bros Burgess and Clutton of King's Cross handed over at Grantham to Bros Marden and Hanson of Leeds, who also found the engine unresponsive all the way.

Three trips with a V2 showed timekeeping, but little margin. Engine 808, with Bros Marshall and Nichols of Grantham, took matters slowly at the start and had a pw at Hatfield so that the first 17½ miles took 30 minutes, leaving 90 minutes for the 88 miles to Grantham. It was done exactly, including a fine climb up the 12 miles to Stoke summit in 13½ minutes, an excellent performance.

Nos 4871, Bros Higgs and Harrison of Doncaster, and 956, Bros Jenkinson and Crouter, on the 9.37am from Grantham with a 118-minute schedule for the 105½ miles, did almost exactly equal performances, and though four pw delays made both trains about 15 minutes late, no time was against loco. I do not often meet Colwick members, but travelled with Bros Horsfall and Crabtree who were working a GC Atlantic 2918 to Nottingham, and had an interesting talk which I much enjoyed.

On the 5pm Paddington to Bath I met Bros Robinson and Pile of Laira. Bro Robinson is an old and enthusiastic member who has done good service for the Society as secretary. They had King class 6013, with 411 tons on a schedule of 117 minutes for 106¾ miles, and were delayed badly by pw and a track circuit failure at the start and two stops outside Bath. Apart from these delays, time was well kept. Returning with Castle class 4091 and 415 tons, Bros Cowan and Rowley of Old Oak – two coaches were slipped at Didcot – the run was spoilt by pw and signal delays from a train ahead, but no time was booked against loco.

All these runs suggest that, in the present conditions, the schedules are more difficult today than the fast timings of 1939.

On the Cornish Riviera and the Scotsman

January 1948

I went to Truro on the famous Cornish Riviera Express. Engine 6004, King George 111, hauling 391 tons, was worked by two Laira members, Bros Hawker and Stephens. Coal was not too good and the engine would not steam as freely as she should have done. Owing to repairs on the line we were sent through Westbury instead of the bypass, and had to stop there seven minutes for water. We were half-an-hour late passing Taunton, but made an unexpectedly good climb up Whiteball, and kept time to Exeter. A stop at Newton Abbott as usual for the pilot, a Bulldog 3401, and we reached Plymouth without more loss, despite a long pw after Hemerdon. Bro Adams of Laira worked the train on with three coaches less and 5022, Wigmore Castle, a rough rider but good steamer which had no trouble with the schedule.

Returning with the same loadings, Bros Goddard and Burt and Castle class 5050 ran into Plymouth dead on time. Here 6004 came on again, this time with Bros Hocking and Rogers of Laira and Bulldog 3991 to Newton Abbott. We were ahead of time at Exeter, but again the engine was not on top form; we went into Frome and stopped for water which, all told, must have cost nearly 15 minutes, and that and pw checks made us 20 minutes out at Paddington.

On the Flying Scotsman, one of the new non-streamlined A2 Pacifics 524, named Herringbone after the 1943 St Leger winner, proved a wonderful steamer and hauled 501 tons over the 76¼ miles to Peterborough in booked time, 88 minutes with three pw slacks costing eight minutes included. Bros Trafford and Manley of Peterborough, and 524, with 134 minutes allowed for just under 112 miles to York, made the schedule look child's play, reaching York to time after waiting nearly five minutes outside. Here an older A3,72, took on, and Bros Newton and Parker of Gateshead bettered schedule to arrive at Newcastle two minutes early.

Returning with 450 tons and A3 69, Bros Ross and Parker of Heaton did excellently to Darlington, recovering a five minutes late start from Newcastle, and then Bros Buckborough and Howe of Darlingtom took over to Grantham. But after York the train ahead blocked us all the way, and after Doncaster the engine, fed with the poorer coal from the bottom of the tender, began to falter and gave Bro Howe work to keep steam up, though no more than two minutes could be booked against loco. From Grantham a similar engine, 83, Bros Parr and Selby of Grantham, went on,and pulled back a minute or two so that arrival was only 15 minutes out.

Nationalisation and standard types

February 1948

In the future even our largest express engines will be too small for the loads and speeds that will be needed. It ought to be possible to put on British rails an engine capable of hauling at least 500, possibly 600, tons at 60 mph start to stop, with a margin for the recovery of the incidental delays which every long run meets. That would require a 4-8-2 or even 4-8-4 in a country where there is not now a single eight-coupled express engine running. In America and some of the Dominions the heavy fast work falls to such engines as these, and the number of wheels makes it possible to keep the weight distributed within the limits set by the road.

The first snag is in turntables. Engineers in the past have been thwarted again and again by the fact that really big machines mean rebuilding turntables all over the line. It has been proved that wheels of 6ft 6in are capable of really high speeds, and with better balancing, exhaust ports and valve gear, there is no reason why we must go over the 6ft 6in mark for speeds of up to 100 miles an hour.

An enlarged edition of the LNER or Southern Pacifics would provide such an engine. For a general utility my vote would be the Stanier Black 5, and the Stanier 2-6-4 for local work. For a four-wheeled couple engine, for which there is still a place, the Southern Schools have no rival. For freight traffic with average loads, I think the GWR 2800 type, though old, still the best, and the LNER J11 freights, rebuilt from the famous old Robinson 0-6-0s of the Great Central, maybe the best of that type, though the ugly duckling C class of the Southern have claims.

Perhaps the diesel will sweep the board as regards many types of traffic. I hope not, for the real skill of a driver and fireman's job will depart with these glorified motor cars. I have often watched real artistry on the footplate in the handling of steam engines. I suppose all that will be needed in future will be to pull a handle and watch the signals!

Shunting tanks are long past their best and poorly kept

232 High Street
Eastleigh, Hants

March 1 1948

Dear Mr Waterhouse

Your article in the February 1948 *Locomotive Journal*, prompts me to write to you as a very disgruntled shunting driver who, owing to a failure to reach the required eyesight standard, is confined to shunting but who is nevertheless interested in the art of engine driving.

My chief bone of contention is with those old six-wheeled coupled tanks which do the brunt of the work in the marshalling yards in this region. Working 24 hours a day for six days a week, these veterans are long past their best, although if kept in a good state of repair still capable of doing the job. But I am afraid that they are not afforded regular and thorough visits to the shops, and thus you can imagine the condition of these cinderellas of the motive power department.

I received a complaint from my shed superintendent regarding the condition of one of these engines. The big ends and the slide blocks knocked so loudly that it reminded me of the clippety-clop of the old grey mare. At the same time the spindle glands were blowing with the regulator worked hard.

Raising the question at a branch meeting I was informed that these engines were going through the shops as soon as possible, but I'm afraid the Battle of Britain class takes priority when in this respect the "Battle of the Shunting" class is equally important.

I read that two of the other regions were ordering new shunting engines. Let's hope the Southern Region will soon follow suit and give shunting drivers a really good tool with which to do the job.

With best wishes
Yours faithfully
Reg Smithwell

The blessings of the straight and narrow

27 Clitterhouse Crescent
Cricklewood
London NW2

March 29 1948

Dear Mr Waterhouse

Thanks for your letter, and I'm glad you like the story of 1294. As you say, she is now a ghost, and it did not cause many tears when she went the way of all scrap iron. Ever since she was stationed at Cricklewood, about 1920, she was nothing but a pest.

She used to work the Radlett Flyer, three coaches, and it is said that the people of Mill Hill, Elstree and Radlett used to run a sweepstake on how many times she would stick for steam. Still, she could move, as one night I was on her and she did Pancras- Hendon in eleven minutes and XL timing was ten mins thirty secs.

As to how she got to Shap via Kettering, I expect it was the same way that some years ago a retiring Kentish Town driver got from Balmoral to St Pancras non-stop when he claimed to have driven the world record non-stop train using a single wheeler, i.e. across the fields.

If you can get a trip on a Garrett you will find it an interesting experience. I think you railway enthusiasts neglect this side of the picture too much.

With regard to GW 28s, on Wednesday I have to attend Euston Station to discuss with the heads the interchange of engines, and I understand that the 28s are going to be tried out on our Luton trains. This will prove how good they are.

Needless to say I'm keen to try a Liz out on the Western, and I think one of these would solve their West of England problems on expresses.

I've seen that animated coach 10 000 several times lately, and so far I'm not unduly impressed.

My opinion of the Black 5s is the same as yours, that is, outstanding. Also I think that Western won't have an easy task to beat a 5X either for speed or acceleration.

If possible I will make my way to Richmond on the Velocipede one day, and perhaps bring Eric, the bloke who makes things screwy every time we go out together. It appears my niece used to be a typist at your college at Richmond, so perhaps she knows you.

By the way, why are so many clergymen interested in railway engines? I've often tried to puzzle this out. The only conclusion I can come to is that engines and trains are a good example of the blessings of the straight and narrow path, and they illustrate very convincingly what happens when they leave it, in other words when they land in old England.

Torag Beg is coming this way in April with his daughter Mary and I am taking them to the opera, having made him into an opera fan on his last visit. His daughter already is one, but she couldn't persuade pa to go.

Yours sincerely
C Brooks

Every company has its own favourite engine

67 Thurlstone Road
Penistone
Nr Sheffield

April 12 1948

Let me first thank you for your interesting article each month in the *Journal.* I had better introduce myself before going further. I started on the Great Northern Railway in 1910 at Barnsley, transferred to Penistone 1915, appointed fireman in February 1919, passed fireman 1924, transferred to Great Eastern section at March (Norfolk) in 1931, returned to Barnsley owing to my wife's health in 1944, and here I hope to stay.

In trade union matters I've always been interested. Our long lost but brilliant Bro Arthur Mason started my interest way back in 1916, and from there I was branch collector and check steward to 1926; branch secretary 1926-31; 1932 elected vice-chairman for all my stay there with the exception of the last two years, I believe. I am not boasting about my union activities but thought it might interest you in a confidential way.

Your recent comments on "Which are the best engines?" interest me, for I can claim to have tried quite a few. Like the question whose is the bonniest baby? Why, every mother has it, and I believe it is the same with engines; every company has it, or in many cases the men think so. I think this is more typical of GE men than any. I used to tell them that if they put a GE man on a wheelbarrow she would be a good one. Mind you, I don't altogether disagree with them.

For comfort and clean working give me a Swedey. In the GE section there was nothing to beat one of those Class H 0-6-0 type goods. What they would be like in South Yorks I don't know. This can also apply to the Clanders, especially the rebuilt one. I had a lot of good runs from March to Norwich and back. The best mineral engine is undoubtedly the Robinson ROD type, there's none better for simplicity and all round adaptability. The Gresley 2-8-0 could leave them on the London line I'll admit, but by far and large give me a ROD.

The best all round in our group is a J39. These are very good engines and I have worked all kinds of trains with them. In fact we had in our link at March the 9.10am London and back each night and they were indeed a success. The K3 is another good one but apt to get very rough when run down. One very interesting trip I had was a troop train from March to Doncaster and Grimsby Docks then back to March via Boston with engine 2H53 all the way without cleaning the fire and only oiled the inside big and small end, a distance of well over 200 miles.

'We have lost pride in our calling'

One regrets the passing of the old engines. The finest one was certainly the GE Atlantic; although there was no comfort, these were a wonderful type. I think also engines have been scrapped sort of willy-nilly, for I'm convinced there was years of work left in many of them. For instance, converting the GE 0-8-0 type into tank engines, well there were 25 good engines lost, for when superheated those old engines could do a good job of work.

I think that the old enginemen of the past were enginemen and no mistake. Take Leicester men with a GE Atlantic, they could get as good results as anyone, or the late Driver Armstrong of Gorton and his 5195. He must have done thousands of miles with her.

In closing, there is one thing. We have lost pride in our calling. Take the state of footplates in general. They are a disgrace to us, for it is surprising what ten minutes' wiping will do.

Best wishes
Yours Fraternally
Bernard Shaw

BRITISH RAILWAYS

V. M. BARRINGTON-WARD

Telephone
PADDINGTON 1601 Ext. 121

Telegrams
RAEX, NORWEST, LONDON

THE RAILWAY EXECUTIVE
222 MARYLEBONE ROAD
LONDON N.W.1

My Ref: O/MP/8552/2

6th May, 1949.

Rev. Professor E. S. Waterhouse,
The Principal's House,
Richmond College,
Richmond,
SURREY.

Dear Dr. Waterhouse,

Your letter to Mr. Allen of the 3rd May requesting his help in obtaining permission to ride on the footplates of the engines working the 12.15 p.m. Marylebone to Sheffield, on the 7th instant, and the 7.40 a.m. Sheffield to Marylebone, on the 9th instant, was passed to me for attention.

I have pleasure in enclosing two engine passes to enable you to travel on the engines of the two trains.

For your information, the working this week-end is somewhat upset owing to the Rugby Final at Wembley. In both cases the engines will be changed at Leicester , the 12.15 p.m. being worked by an A.3 class from Marylebone to Leicester, and thence by a B.1 from Leicester to Sheffield, and the 7.40 a.m. by an A.3 from Sheffield to Leicester, and another A.3 from Leicester to Marylebone.

Yours sincerely,
for V.M. Barrington-Ward.

Enc.

P.S. Kindly note conditions on back of pass.

See, overleaf, W.P.Allen's letter

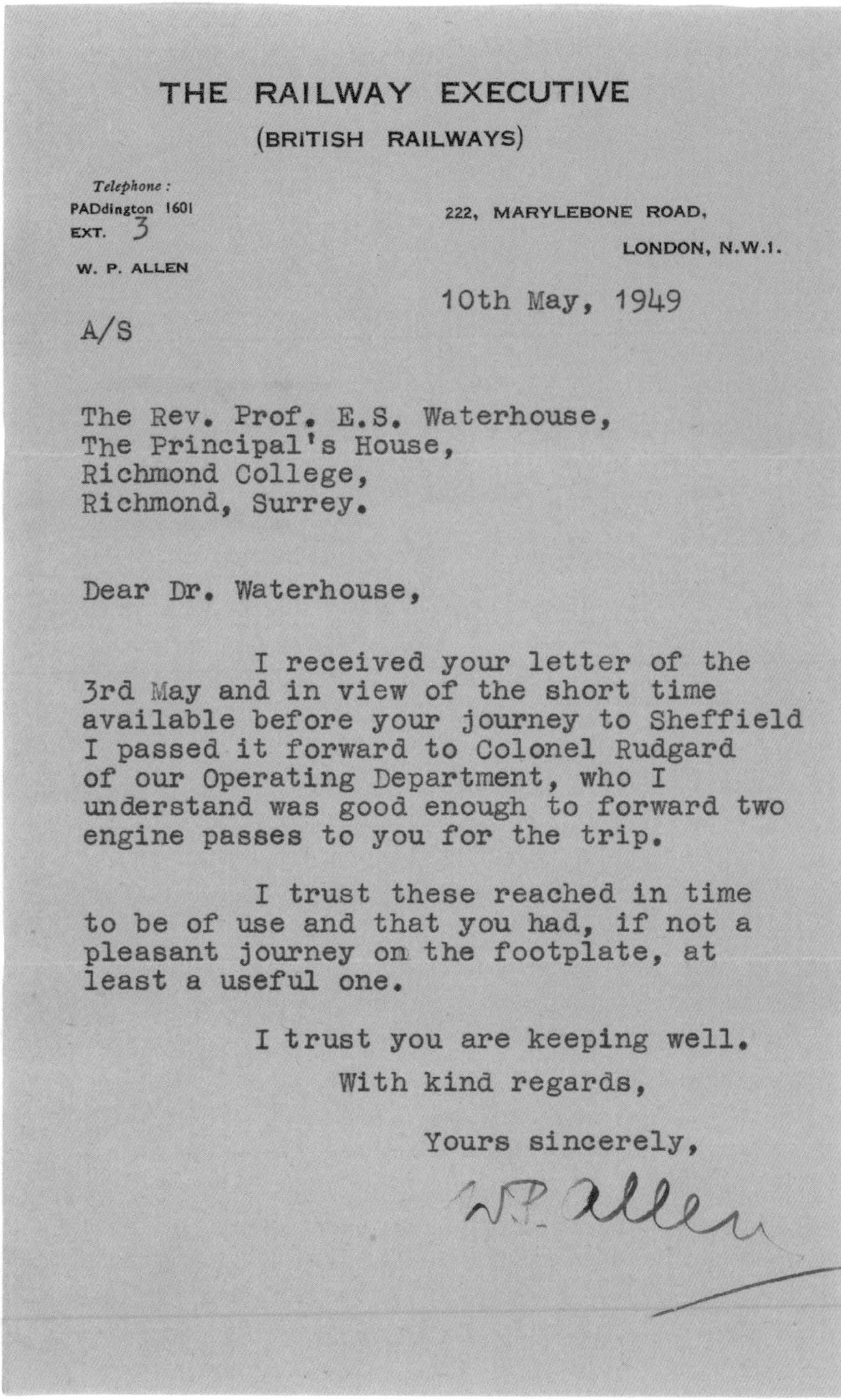

THE RAILWAY EXECUTIVE

(BRITISH RAILWAYS)

Telephone :
PADdington 1601
EXT. 3

W. P. ALLEN

222, MARYLEBONE ROAD,
LONDON, N.W.1.

10th May, 1949

A/S

The Rev. Prof. E.S. Waterhouse,
The Principal's House,
Richmond College,
Richmond, Surrey.

Dear Dr. Waterhouse,

I received your letter of the 3rd May and in view of the short time available before your journey to Sheffield I passed it forward to Colonel Rudgard of our Operating Department, who I understand was good enough to forward two engine passes to you for the trip.

I trust these reached in time to be of use and that you had, if not a pleasant journey on the footplate, at least a useful one.

I trust you are keeping well.

With kind regards,

Yours sincerely,

W.P. Allen

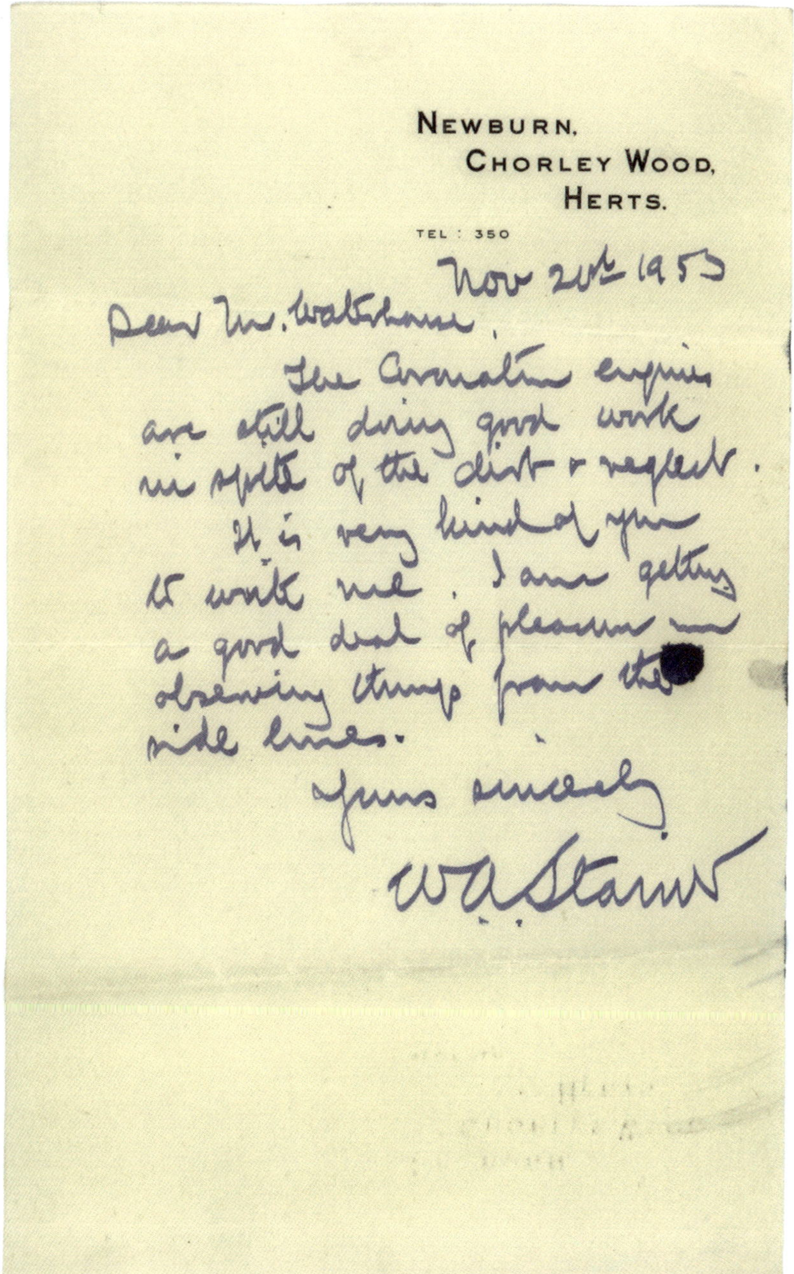

NEWBURN,
CHORLEY WOOD,
HERTS.

TEL : 350

Nov 20th 1953

Dear Mr. Waterhouse,

The Coronation engines are still doing good work in spite of the dirt & neglect.

It is very kind of you to write me. I am getting a good deal of pleasure in observing things from the side lines.

Yours sincerely

W.A.Stanier

On nationalisation, W.P.Allen, ASLEF general secretary, moved across to become a key member of the British Transport Commission's Railway Executive. A Peppercorn class AI locomotive was named after him. He and ESW knew each other well from his time editing the *Journal*. The letter from W.A.Stanier in retirement suggests some sort of friendship. Stanier was, of course, ESW's preferred locomotive designer

Addressing locomotive inspectors

March 1955

By the invitation of Mr E. D. Trask, Motive Power Superintendent, Eastern Region, I had the privilege of addressing a Motive Power Conference (Locomotive Inspectors) at the British Railways Staff Training School, St Ronan's, Hadley Wood, on December 30 last.

I am accustomed to lecturing – it has been my job – but I confess I would rather have addressed the whole bench of bishops than this audience, for whatever I might say they knew more than I did about it. However, they were very kindly and uncritical, and I enjoyed myself. I met many friends, with several of whom I have ridden on the footplate – such as Messrs Hulett, Dolby, Dixon and Emerson.

The hour-and-a-half was equally divided between my talk and the discussion, which would have lasted longer had time permitted. One odd point was the belief that seems to linger that the letters C and T which mark the beginnings and ends of speed restrictions stand for Caution and Through rather than Commence and Terminate.

I was questioned on my remarks about V2 engines. I admitted the excellent work they had done, but did not consider engines with a leading pony truck were suitable for really high speeds as they do not exert sufficient guiding force and tend to sway on the spring links, especially if there is no spring control.

Nobody disputed my notion that engines have individual characters for which no mechanical explanation can be given, though no doubt there is one. That is why, as every engineman knows, there are good ones and bad ones, and they normally remain right throughout their career true to character.

It also seemed agreed that the standard types will not cover the needs of our railways. We have yet to see what Clan Class 72000 is capable of, but it is evident that the Britannias, which were built as mixed traffic engines (and have their points) are not capable of heavy high-speed work or of taking a super-speed lightweight train like The Bristolian.

Nobody seemed able to tell me what I have long wondered, and that is why the stroke of British locomotives is always an even number of inches, for example 26, 28, 30 – never 25, 27, 29. Is it just custom? There is plenty of variety in the bore.

With this unsolved conundrum we broke off and an excellent dinner ended a most enjoyable time.

Steam's swan-song needs the old guard

April 1955

The scheme for the future of the railways sounds like the swan-song of steam, though one may doubt whether it will come as soon as many seem to think; but the intention to stop building more steam locomotives necessarily involves the end sooner or later. Meanwhile for the really fast work we must rely on the old guard, particularly the Midland Coronations, the Southern Merchant Navies, the Western Kings and the Eastern A4s, aided by Scots and Patriots as rebuilt, West Countries, Castles and the A1, 2 and 3 types respectively.

I have not had any direct experience of the new prototype Duke of Gloucester, but the Britannias as far as my knowledge indicates are good all-round engines with a turn of speed, if not suited for heavy express duties, for which indeed they are not designed.

It is probable that the need for more diesels in the yards, and for more suburban electrification, will keep the express engine at work for many years yet. Anyhow, it is likely to last me out and that's a comfort, for I can get no enthusiasm for the other types of locomotion!

As an example of heavy fast work, I return to the Mid-Day Scot of which I witnessed a remarkable run recently. This time the engine was 46249 City of Sheffield with Bros Speed and C Russell of Crewe, accompanied by Inspector Smith of Camden. The load was 402 tons. The timing to Watford is a bit optimistic for such a load, and we took 23 instead of 21 minutes for 17½ miles, but recovered one by Bletchley and then romped along, doing the next 36 miles in 31 minutes to stop at Rugby, 82½ miles in 79 minutes, one minute early.

Leaving Rugby we began to get a bit ahead in view of the pw slack at Betley Road, but we lost two minutes to signals at Colwich, and then were stopped dead at Great Bridgeford, exactly as on a former trip, by a Stoke connection of the 12 o'clock from Euston. This put us down by five minutes, but the actual running was all that could be desired.

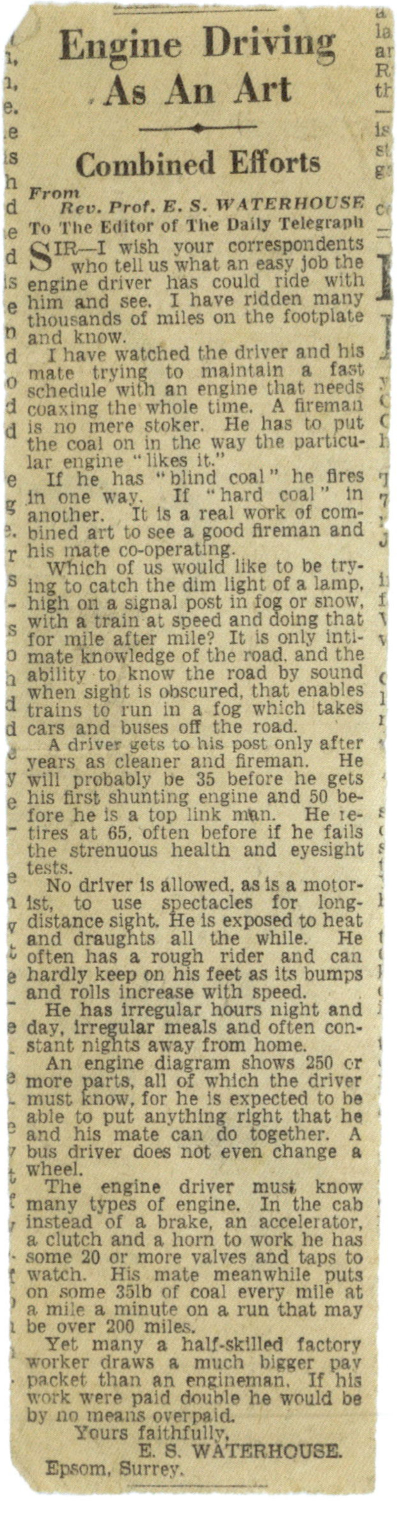

Engine Driving As An Art

Combined Efforts

From
Rev. Prof. E. S. WATERHOUSE
To The Editor of The Daily Telegraph

SIR—I wish your correspondents who tell us what an easy job the engine driver has could ride with him and see. I have ridden many thousands of miles on the footplate and know.

I have watched the driver and his mate trying to maintain a fast schedule with an engine that needs coaxing the whole time. A fireman is no mere stoker. He has to put the coal on in the way the particular engine "likes it."

If he has "blind coal" he fires in one way. If "hard coal" in another. It is a real work of combined art to see a good fireman and his mate co-operating.

Which of us would like to be trying to catch the dim light of a lamp, high on a signal post in fog or snow, with a train at speed and doing that for mile after mile? It is only intimate knowledge of the road, and the ability to know the road by sound when sight is obscured, that enables trains to run in a fog which takes cars and buses off the road.

A driver gets to his post only after years as cleaner and fireman. He will probably be 35 before he gets his first shunting engine and 50 before he is a top link man. He retires at 65, often before if he fails the strenuous health and eyesight tests.

No driver is allowed, as is a motorist, to use spectacles for long-distance sight. He is exposed to heat and draughts all the while. He often has a rough rider and can hardly keep on his feet as its bumps and rolls increase with speed.

He has irregular hours night and day, irregular meals and often constant nights away from home.

An engine diagram shows 250 or more parts, all of which the driver must know, for he is expected to be able to put anything right that he and his mate can do together. A bus driver does not even change a wheel.

The engine driver must know many types of engine. In the cab instead of a brake, an accelerator, a clutch and a horn to work he has some 20 or more valves and taps to watch. His mate meanwhile puts on some 35lb of coal every mile at a mile a minute on a run that may be over 200 miles.

Yet many a half-skilled factory worker draws a much bigger pay packet than an engineman. If his work were paid double he would be by no means overpaid.

Yours faithfully,
E. S. WATERHOUSE.
Epsom, Surrey.

The ASLEF drivers' strike of May-June 1955: ESW's letter to the *Daily Telegraph*

Such was the importance of rail transport to the nation that a State of Emergency was called to keep Britain moving when ASLEF, partly joined by the NUR, began a rail strike on May 28 1955. The dispute over pay differentials resulted in increases being awarded to engine drivers (but not firemen) and was called off on June 14.

The letter from ESW, left, was published in the *Daily Telegraph* of May 10 1955, more than a couple of weeks before strike action. It's shown in page context opposite, perhaps surprisingly accorded lead letter status given that the *Telegraph* and most of the media had little sympathy for the strike.

Two letters from the archive on page 120 record a positive reaction to ESW's viewpoint; the one from an ASLEF branch official is more predictable

DAILY TELEGRAPH
AND
MORNING POST
TELEGRAPH - June 29, 1855
Post - November 2, 1772
[Amal]gamated October 1, 1937
[TUE]SDAY, MAY 10, 1955
[Fle]et Street, London, E.C.4
[Tele]phone: Central 4242

[POLI]CY CONFLICT IN THE EAST

[...] from planning talks [...] Russia, the Foreign [...]' chief task in Paris is [...] to hammer out an [...] practical and consistent [...] Indo-China. At present [...] confusion cloaks the [...] policies being pursued [...] ce and America. This [...] concealed the extent [...] they are incompatible [...] another.

[...] policy is in itself in a [...] some confusion. It [...] to retain a lingering [...] for the Emperor BAO [...] regarded in Paris as [...] man who could unite [Vi]et-nam. It has certain [...] h the dissident sects [wh]ich were France's allies [...] war against Viet-minh [...] ch are understood to [rece]ived the aid at least of [...] ate French officials in [str]uggle against Mr. DIEM. [...] is any premise common [...] French thinking about [...] na, it is the conviction [...] DIEM is a reed about [...] oken—a conviction that [...] nt successes will do [...] o remove. The French [...] assume that the [el]ections promised by the [...] agreement will be held [...] he premature collapse [...] EM renders them un- [...], that Ho CHI-MINH [...] p all before him, and [n]othing consequently [...] to be done except to [...] om him the best terms.

At this very moment, [n]egotiators in Hanoi are [...] safeguard French com- [...] and cultural interests [...]

[...] an policy is based on [...] ary assumption that [...] is far from a broken [...] that South Viet-nam [...] ade a going concern, [...] aid of American cash, [...] advice and prestige. [...] hardly deny the [...] esis that, if elections [...] Ho will win them; with [...] h regimented and [...] ly superior, and the [...] merically inferior and [...] no other conclusion is [...] The fact that the [...] would not be "free" in [...] ense of the word would [...] mericans an excuse to [...] r. DIEM not to hold [...] vice which would [...] be taken.

[...] part, the British are [...] only to ensuring [...]" elections take place. [...] words, we are in honour [...] a course which, unless [...] ion changes dramatic- [...] not only embroil us [...] Americans; it will also [...] handing over South [...] rulers, refugees and [...] ke, gagged and bound [...] unist vengeance. We [...] ive our consciences by [...] that this honourable [illegible] [...] e conclusion.

[...] a tangle have we [...] can be unravelled in [...] If we share the view [...] DIEM is finished, then it [...] siness to persuade the [...] of the folly of attach- [...] prestige to his. If we [...] en it is our disagree- [...] to help prolong the [...] rtition (which the im- [...] of holding free [...] permits us honourably [...] might also point out [...] ench the futility of [...] the Americans by [...] to negotiate with Ho. [...] have no intention of [...] s word in any case; so [...] r. DIEM survives with [...] support in the south [...] od excuse not to do so.

[...]L TROUBLES

[...] was an improvement [...] rather than a de- [...] in the Yorkshire [...] or the first time since [...] ouble developed early [...] k. Only moderate [...] n can be derived from [...] Great damage has [...] en done to the nation [...] g the gap between the [...] coal needed for indus- [...] other uses and the [...] oduced. Worse still is [...] ent revelation that [...] es in some of the [...] re so high that we are [...] to get either peace or [...] production.

[...] ins of this particular [...] not recent. Broadly [...] they are the long- [...] ifficulties in fixing the [...] ion of piece-workers [...] parts of their jobs [...] ot be quantitatively [...] But what is disquieting [...] trouble should have [...] at the moment when, [...] st, the National Coal [...] the union had found [...] hich, after extensive [...] increased both earn- [...] output; and that the [...] hould have won the [...] more than half the [...] the coalfield. The [...] are that the relations [...] nen and management [...] level are still highly [...] tory; and that the [...] of the union is far [...] trouble than when it is out for peace.

This second feature is not confined to the mining industry: and though it may be a matter of human nature that obedience is rendered more readily in war than in peace, responsible trade union leaders are quite right to be alarmed. In the case of the mining industry, the wage-earners have been promoted with general assent from the Cinderellas to the aristocrats of labour. The improvement in status involves an increase in responsibilities, the very first of which is not to inflict damage on the nation by an easy recourse to strikes. In the present case there is substance in some of the men's complaints: but surely there is none in the particular complaint that the machinery of conciliation is so ample that any strike must be unofficial. Ample machinery should be amply used.

SALK VACCINE

BRITAIN's decision to put off tests on children with the new poliomyelitis vaccine has been amply justified by the subsequent crisis in the American immunisation programme. Inoculation there has been temporarily stopped after more than 5m. children have received it, and 54 of them have contracted the disease against which they were supposedly being protected. This does not warrant withdrawal of faith in Dr. SALK's discovery, still less a panicky conclusion that 5m. children have been placed at risk. The evidence does suggest, however, that this nation-wide programme of inoculation was launched before the problems of controlling the manufacture of the vaccine on a large scale had been fully worked out.

Development of any new life-saving drug, vaccine or serum must almost inevitably present a crisis of decision. Short of the 100 per cent certainty that is beyond human attainment, there comes a moment when the value of the remedy or immuniser seems established by laboratory and field tests. Then the doctors have to decide between making it generally available, in the certainty that many lives will be saved—or holding it back for yet further checks, in the fear that some healthy lives may be risked. There can be no rule of thumb for making such an agonising decision: but caution should always be the guiding principle, most especially in vaccination. The caution shown by the British authorities will enable us to benefit from the fresh efforts now being made in America to narrow the limits of fallibility.

Engine Driving As An Art

Combined Efforts

From Rev. Prof. E. S. WATERHOUSE
To The Editor of The Daily Telegraph

SIR—I wish your correspondents who tell us what an easy job the engine driver has could ride with him and see. I have ridden many thousands of miles on the footplate and know.

I have watched the driver and his mate trying to maintain a fast schedule with an engine that needs coaxing the whole time. A fireman is no mere stoker. He has to put the coal on in the way the particular engine "likes it."

If he has "blind coal" he fires in one way. If "hard coal" in another. It is a real work of combined art to see a good fireman and his mate co-operating.

Which of us would like to be trying to catch the dim light of a lamp high on a signal post in fog or snow, with a train at speed and doing that for mile after mile? It is only intimate knowledge of the road, and the ability to know the road by sound when sight is obscured, that enables trains to run in a fog which takes cars and buses off the road.

A driver gets to his post only after years as cleaner and fireman. He will probably be 35 before he gets his first shunting engine and 50 before he is a top link man. He re- [illegible] the strenuous health and eyesight tests.

No driver is allowed, as is a motorist, to use spectacles for long-distance sight. He is exposed to heat and draughts all the while. He often has a rough rider and can hardly keep on his feet as its bumps and rolls increase with speed.

He has irregular hours night and day, irregular meals and often constant nights away from home.

An engine diagram shows 250 or more parts, all of which the driver must know, for he is expected to be able to put anything right that he and his mate can do together. A bus driver does not even change a wheel.

The engine driver must know many types of engine. In the cab instead of a brake, an accelerator, a clutch and a horn to work he has some 20 or more valves and taps to watch. His mate meanwhile puts on some 35lb of coal every mile at a mile a minute on a run that may be over 200 miles.

Yet many a half-skilled factory worker draws a much bigger pay packet than an engineman. If his work were paid double he would be by no means overpaid.

Yours faithfully,
E. S. WATERHOUSE.
Epsom, Surrey.

SWEET TOOTH

To The Editor of The Daily Telegraph

SIR—The report of the findings of the Medical Research Council on sugar consumption and dental decay is likely to mislead parents into thinking that sugar and sticky starches are not the major factors in causing disease of the teeth. Whether the amount of sugar taken has any bad effects on the teeth must be left to the experts, but there is no doubt that the main cause of tooth decay is the retention of these substances in the mouth.

Every dentist who has treated young children knows of the extensive decay caused by putting jam or similar substances on the "dummy." Even if it were possible to restrict the child to the small amount of sugar taken in this way, these teeth would still decay.

It is the length of time the sugar and easier fermentable starches are allowed to remain in contact with the teeth that determines the rate of decay. Parents should take the advice of their dentist on the care of the teeth. Yours faithfully,
D. E. TAYLOR.
Principal School Dental Officer.
Leeds.

TO-DAY'S EVENTS

THE QUEEN and Duke of Edinburgh visit B.I.F., Olympia, 11.
St. Paul's Cathedral: Merchant Taylors' School Service, 5.30.
Univ. of London: Wright-Fleming Inst. of Microbiology: Dr. D. A. Long, "Response of Host to Bacterial Infection," 5. Univ. Coll.: Dr. M. Pickford, "Anti-diuresis," 5.15; Prof. H. C. Brown, "A Proposed Quantitative Treatment of Aromatic Substitution," 5.30. Senate House: Prof. J. A. O. Larsen, "Boeotia & Oligarchic Theory of 5th Century," 5.30.

No Frontiers for Coal & Steel

How European Community Works

By E. V. FRANCIS

WHEN the Common Assembly of the European Coal and Steel Community met in Strasbourg yesterday, on the fifth anniversary of the Schuman Plan, the first milestone towards Western economic integration had already been passed. There is now no doubt about the Community's success. Its leading members, France and Germany, have endorsed that success by their agreement to promote integration in other industries.

This new impulse to the movement towards Western unity marks its revival after the severe political setback over the European Defence Community last autumn.

The next stage is still some way ahead. For France and Germany have only agreed in principle to move to integration in the spheres of aircraft industries and atomic development. The other Schuman Plan partners — Belgium, Holland, Italy and Luxembourg—are also involved, and Britain will have to be consulted by virtue of her association with the Community. But these are problems for the future. The main point is that the Coal and Steel Community has survived its initial trials to provide a pattern for the further integration of Western Europe.

Powers and Duties

WHAT is the Community? How does it work? What has it achieved? What sort of pattern does it provide?

Created by the six-nation Treaty of April, 1951, its purpose is the pooling of their coal and steel resources within a common market. Although it owns not a single pit or furnace and depends on the member Governments to give effect to its decisions, the High Authority, as the executive of the Community, exercises at its Luxembourg headquarters very wide powers. It can command the removal of Customs tariffs, the liquidation of cartels if required and the inspection of company records.

The success of the Community owes much to the fact that its functions and constitution were clearly set out from the start. The High Authority is subject to the annual vote of the Common Assembly and to a Court of Justice, which keeps it within its terms of reference. Its Council of Foreign Ministers provides a link with member governments, and its Consultative Committee with the producer, consumer and trade union interests affected by its decisions. In this way, the social political and economic implications of its policy are brought to bear upon the High Authority.

The High Authority's mandate covers the industrial heart of Western Europe in an area of half a million square miles with a population of 160 m. people and a coal and steel output equal to that of Russia and second only to that of the United States. Within this area —now the common market—there is a free movement of coal, iron ore, steel, scrap and special steels regardless of frontiers.

Trade has been stimulated: 75 per cent. more steel was exchanged within the Community in 1954 compared with 1952, and 25 per cent. more coal. Production of steel is running at the record rate of 50 m. tons annually and coal at a post-war peak of 250 m. tons.

This has followed on drastic trade reforms by the High Authority.

Customs tariffs have been eliminated, as well as currency restrictions, quotas and preferential freight rates on national railways. Formerly, a trainload of Lorraine coal travelling to its natural market in south Germany had, for example, to pay Customs duties and railway tolls at the frontier, as well as the higher freight rates on foreign traffic within Germany itself. At the same time the Aachen collieries in the north were forced to compete with Lorraine for this market as a result of their being cut off from their natural market in Belgium and Holland by similar levies and restrictions.

Economic absurdities of the kind are becoming a thing of the past. There are through rates for cross-frontier traffic on all the national railways, and border tolls will be progressively eliminated within two years. Lorraine will get cheaper coke from the Ruhr, Belgium cheaper coal from Lorraine, and so on. Freed from its previous strait-jacket, trade has begun to flow in accordance with the competitive play of the market.

It remains only for the High Authority to do battle with the big cartels, German, French and Belgian, which dominate the trade in coal to remove the last major restraints on competition within the common market.

In this it faces perhaps its greatest test. For these cartels have been built into the economic structures of their countries, and are identified in varying degrees with the interests of trade unions and consumers as well as of producers. Nevertheless, the High Authority is committed to take action under its anti-cartel law, and some less objectionable form of organisation is likely to be imposed on them.

Paying for Progress

INCREASING competition within the Community will cause many uneconomic collieries and steel plants to be closed down, reorganised or merged into larger units. Such rationalisation requires new investment, and as some 20,000 workers are likely to be made redundant in the process, facilities for re-training and re-settlement of labour have to be provided.

With an income of $50m. a year from its levy on the Community's production, the High Authority provides guarantees for loans and contributions to the cost of selected projects required to assist both producers and workers in meeting technical changes of the market.

Of the first $100 m. loan raised by the High Authority in the United States, $75 m. has been allocated to 41 firms to help to modernise plant and to finance pit-head electricity schemes. The other $25 m. has been earmarked to help to build 100,000 miners' houses within five years.

Among the labour schemes is the provision of a bonus of £200 for each miner moving with his family from the declining mines in southern France to the expanding Lorraine coalfield.

Again, the High Authority has contributed equally with the French Government to a 840,000 dollar fund to guarantee a minimum wage and retraining facilities for 1,500 workers affected by the two-year modernisation project of the new steel merger in St. Etienne.

Altogether some $12 m. has been set aside for re-settlement projects. There is also a plan for a European labour card, so that miners and steel workers can take jobs within any of the six countries.

This is a good start. But how far can the High Authority get towards the free movement of labour within the Community? One essential condition is that workers should carry their social security benefits from one country to another. Then again, wage conditions and tax structures must be aligned if the redistribution of labour within the Community is to be assured. Yet the High Authority has no mandate to deal with such matters.

The trade unions are already conscious of these difficulties, which they will shortly be discussing with the employers in the Consultative Committee. But it is doubtful if much progress can be made along these lines, since the decisions lie not with this Community but with the governments.

The Next Step

THIS is not an isolated problem. It is inherent in the High Authority's being confined to a single economic sector which cannot be divorced from the rest. It cannot, for example, plan the Community's coal resources in isolation from electricity and other forms of energy. It has been impelled to help finance electricity plants at collieries in order to use low-grade coal that could not be sold profitably, if at all, in the form of solid fuel. Yet the High Authority has no say in electricity policy.

This underlines the fact that the Schuman Plan intended the Community to be only the first step towards integration. The next step was to have been the European Defence Community. No such link exists with the Western European Union which has replaced it. Nevertheless, the success of the High Authority and the latest Franco-German agreement suggests that there is nothing in the present political framework which inhibits progress towards further integration. And as long as the Bonn Government does not seek unification of Germany at the expense of western co-operation it will continue to be so.

The immediate lesson of the High Authority is that the way to further economic integration lies in having a limited objective in any field and granting clearly defined powers for its achievement. On the present showing, there is considerable leeway to be made up in this sphere before wider political arrangements are likely to become imperative.

LONDON DAY BY DAY

At Wellington's Oval Desk: Royal Colonels-in-Chief

IN a room which has changed little since it was built almost exactly 200 years ago, I spent yesterday morning with Major-Gen. George Johnson, Major-General commanding the Household Brigade.

As chairman of this year's Royal Tournament at Earl's Court, the 75th to be held, he had just announced details of this popular June event.

It is from Gen. Johnson's room, immediately over the arch leading from the Horse Guards Parade to Whitehall, that the Queen Mother will watch the Trooping on June 9.

Also from here the Major-General, a distinguished ornithologist, can spend his leisure moments watching the duck in St. James's Park.

Election Forecast

A study in white and green designed by Kent, the room has but [illegible] son sits at his desk he cannot avoid the basilisk stare of the Duke of York, perched on his column a few hundred yards away.

The oval writing desk belonged to the Duke of Wellington when Commander-in-Chief. It has four knee holes so that the Duke's staff officers could overcome his deafness in conference.

The list of those who will be in the Royal box at the Tournament contains, I notice, an intelligent political forecast. On June 10 it will be Mr. Attlee, "Leader of the Opposition."

Perhaps it is as well that Parliament is not sitting. Otherwise Mr. Woodrow Wyatt, that persistent inquisitor of the Brigade of Guards, would surely raise the matter on the adjournment.

Diminished Galaxy

KING FREDERIK'S visit yesterday to his regiment the Buffs, prompted me to look through the records of other foreign Sovereigns who have a similar honour.

At present there are only two. The Green Howards have King Haakon as their Colonel-in-Chief, and the Royal Sussex Queen Juliana.

As one might expect this compares with a far brighter galaxy earlier in the century. The Emperor Franz-Joseph, the Kaiser, the Czar and Alfonso XIII of Spain were all colonels-in-chief of cavalry regiments.

King Haakon was then honorary colonel of the Norfolk Yeomanry.

Admiral Since 1905

Honorary admirals are almost as scarce as colonels-in-chief.

They consist of the three Scandinavian kings—King Haakon's appointment, incidentally, dates back to 1905—and King Paul of Greece.

King Feisal of Irak was appointed an air vice-marshal the other day. Prince Bernhard is an air commodore.

Diver Forbidden to Swim

CAPT. JACQUES COUSTEAU tells a diving story in his introduction to "Two Thousand Fathoms Down." This account of the record descent off Dakar by Cdr. Houot and Lieut. Willm, of the French Navy, appears next week.

My picture shows Willm in the conning tower of the bathyscaphe, their deep-water submarine.

Pierre Willm at the hatch of the bathyscaphe

When the time came for Cousteau to give up his command of the research ship he met his successor, Houot, for the first time and explained his duties.

Houot's face fell. Following an attack of poliomyelitis he had been forbidden sea bathing. If he could not bathe, how could he dive?

Fortunately the anomaly of replacing Cousteau by an officer who could not get wet proved more [illegible]. The bathyscaphe went down to greater depths than man has ever reached before, but Houot and Willm inside it remained perfectly dry.

Time and Gold

THE superb setting in front of Buckingham Palace is now being completed by the gold leafing of Australia Gate.

Canada Gate, on the Piccadilly side, received its gold leaf during the winter. It forms a striking background to the tulips, which are now at their best.

It took four months to gild Canada Gate. The work on Australia Gate is to take three weeks.

The Ministry of Works says the weather was largely responsible for this discrepancy between the Dominions. One wonders if the winter can have been quite that bad.

Queen Mother's Friend

LADY ANNALY, who has died after a long illness, will be remembered for her gaiety, her wit and above all her kindliness.

Born a Spencer and the sister of the present earl, she inherited the beauty and liveliness which Joshua Reynolds recorded of her forbears two centuries ago.

For generations her family has been closely associated with the Court. It was the more natural that she should follow in the tradition since Queen Elizabeth the Queen Mother was her friend from girlhood.

The Queen Mother called on Lady Annaly at the Westminster Hospital within two hours of her return from America.

Maurice Baring's Niece

Lady Annaly shone as a hostess. She entertained all her guests, irrespective of station and distinction, with the same grace and vitality.

And when in the mood her verbal felicities with their clear-cut precision reminded one of her uncle, Maurice Baring.

Bond Street Prepares

A GENERATION and more ago Ernest Rechnitzer, who then lived in Mayfair and was a man of a certain note in the City, collected furniture.

Now in old age Mr. Rechnitzer, a merchant banker of Hungarian extraction, lives in Brighton, and he has decided that at least some of the proceeds of his connoisseurship shall be sold.

The sale, at Christie's on Thursday week, will be a field day for the Bond Street dealers, who are preparing for high prices.

Nearly all the furniture is 18th-century French. Some of it is of the highest quality.

From Rifle to Needle

THOSE who regard needlework as a wholly female occupation may correct their error by visiting the Embroiderers' Guild Exhibition. It opened yesterday at the R.W.S. Galleries in Conduit Street.

[...] masculine modesty. The proportion of men to women among the Guild's 2,500 members is very much higher.

Embroidery, I am told, is popular among Servicemen. They find spiritual release in this somewhat radical change from square-bashing.

The Guild owes much to Royal support. Queen Mary, an expert in the art, was its patron, and several of her gifts are on show. The Princess Royal became patron last year.

Talking Against Himself

N.A.T.O. was set a new language problem with the admission of Germany. At yesterday's ceremony I hear it was not successfully resolved.

Dr. Adenauer spoke in German without a microphone. Simultaneously his speech was coming over loudspeakers in English. This meant that he had his own words shouting him down in another language.

[illegible] gave French or German texts of his speech. My informant, who sat through the hour of complimentary addresses, tells me the occasion was saved by the excellent printed texts available.

Nobody needed to listen to anything, although he saw all the Foreign Ministers making courteous play with their earphones.

Figures That Mislead

A STATISTICIAN gave me a warning yesterday about the deductions Conservatives and Socialists will draw from this week's local polls. Mere gains and losses, he says, mislead. What matters is proportions.

Of the 567 urban districts, the Socialists now control 187—far fewer than one might expect from their Parliamentary strength. They hold 117 out of 396 boroughs.

In the recent county council elections the Socialists, controlling 12 out of 62, lost five. That is, half their supremacy vanished.

Thus, further Socialist losses will be a much more serious matter for them than a similar number would be for the Tories.

Jumpy Nehru

ONE of the three Negroes in Congress, Adam Clayton Powell, has had harsh things to say about Mr. Nehru on his return from the Bandoeng conference.

Mr. Powell, who went as an unofficial American observer, was not impressed by Nehru's personal appearance and deportment. He describes him as jumpy, touchy, sensitive and unable to keep still.

Once, he says, Nehru came into the conference room before a meeting, examined all the place cards of the seats reserved for the Colombo Powers, fidgeted with the ashtrays and then tapped each microphone to see that it was working.

Port and Cricket

FOR the second time since the war an English cricket side is visiting Portugal. On Saturday a team drawn from the wine trade sails to play three matches against the Oporto C.C., which is 100 years old this year.

Tom Pearce leads the team. He went out with that of 1952, when he was captain of Essex. Supporting him will be such county figures as W. E. Hazelton, F. St. G. Unwin and Cedric Thistleton-Smith.

Although recruiting is not so easy in Oporto as in the great days of the port trade, the local club hopes to do much better than in its first match against visitors in 1869.

Then the Channel Fleet replied with 205 runs to Oporto's nine in the first innings.

Summer Wines

FROM a wall of the East India and Sports Club in St. James's Square, Queen Victoria yesterday gazed down on a tasting of German wines arranged by John Harvey's [...] because the wines included a magnificent Niersteiner Rehbach Beerenauslese Cabinet 1953—at the equally magnificent price of 480s a dozen.

Of the Palatinate vintages I admired a Forster Ungeheuer Riesling Auslese, smelling of lime blossom, at 333s 6d a dozen.

In general I found the '53 Moselles rather disappointing—dry, with little nose. But of the 1952 Rheingau wines I recommend an aromatic Rüdesheimer Burgweg Riesling Spätlese.

Questioned Marks

MY recent picture of a Sun Insurance mark has brought several letters from experts in this little-known collectors' field.

All agreed that the policy number was much later than 1710. It was, in fact, issued in 1777 to a Rev. William Cole, of Bedfordshire.

The type of mark I illustrated was in use from 1710 to 1807. Variations were 12 rays instead of 16, and they might be wavy or straight.

The Fire Mark Circle reassures me on the subject of rival brigades fiddling while homes burnt. The practice grew up of insurance companies paying the expenses of others which helped.

At Hodgson's last week 33 fire marks of different companies fetched £93.

Landslide Ahead?

IN his election address the Labour candidate for the Castle Bar ward of Ealing urges the electors to "Vote Early, Vote [...]"

Lighting-up time 9.36 p.m. Sun rises 5.18 a.m.; sets 8.36 p.m. Moon sets 7.13 a.m. High water [...]

War Pensions Red Herring

Labour's Election Gambit

To the Editor of The Daily Telegraph

SIR—Before Parliament was dissolved, Mr. C. J. Simmons, the Labour Member for Brierley Hill, saw fit to make the only bitter party political speech on war pensions that we had either in this or the preceding Parliament. Members on both sides of the House have co-operated in keeping war pensions as far as possible out of party politics.

But the General Election's approach was too much for Mr. Simmons, and he well and truly dragged the subject down into the mud. He was critical of everything this Government have done with regard to war pensions, and did not even mention that they have given the two biggest increases in war disability pensions. He then had the effrontery to say he did not want to see the disabled man the shuttlecock of party warfare.

Sir Ian Fraser, President of the British Legion, who followed, had last December given full credit to the Government for the best-ever increases they have made, and he pleaded that during the election we should see whether we could avoid following each other round making speeches about ex-Servicemen and war pensions. As Mr. Simmons's speech was a piece of pure electioneering, it will, I am sure, be condemned by ex-Servicemen all over the country.

There are two points I would like to make. There was an important debate on war pensions in February, 1951, during the last year of the Labour Government. It was an all-party debate, moved by a Conservative Member and seconded by a Labour Member. The Labour Government were strongly pressed to give a substantial rise in the basic pension which had been increased from 40s to 45s in 1946, but, owing to the steep rise in the cost of living under Labour rule, had fallen in value to 35s.

In the Budget that followed shortly afterwards, the Labour Government found themselves unable to do anything more to increase the basic pension rates for the war disabled, and it was left to this Conservative Government in 1952 to relieve their hardship by first giving a 10s increase in the basic rate and 7s to widows, and then giving an even bigger increase last December in almost every form of war disability pension.

Mr. Simmons boasted of the amount of money the Labour Government had spent on the war pensioner. In actual fact, they spent in all sums amounting to £11½ million a year during their six years in office. The present Conservative Government in only three and a half years spent £25 million in pensions increases.

I hope with Sir Ian Fraser that we are not going to see the war disabled made the plaything of party politics, despite Mr. Simmons's speech. I am certain that the ex-Service community will regard this outburst as inexcusable from one who was himself a Parliamentary Secretary in the Ministry of Pensions.

Yours faithfully,
Gillingham. F. A. BURDEN.

**In the Parliament now dissolved, Mr. Burden was Conservative Member for Gillingham.

LIBERALISM'S NEED

From Lord REA
To the Editor of The Daily Telegraph

SIR—The Liberal vote is more important than ever before. The two major parties are locked together in a struggle for power. Power is not government. Liberals do not seek power, but sincerely to ensure that the voice of independence shall be heard in the House of Commons and that Liberal thought and ideas shall keep watch over the precious liberties of the British people—liberties apt to be forgotten in the midst of so bitter a conflict.

Liberal opinion must be maintained, and I therefore ask all Liberals, and all who sympathise with Liberal thought, to back their opinions by subscribing to the Liberal Party Election Fund, 58, Victoria Street, S.W.1.

Yours faithfully,
REA, President.
Liberal Party Organisation.
London, S.W.1.

AUSTRALIAN LEFT

SIR—In your leading article, "Australian 'Bevans,'" you discuss the present split in the Australian Labour party. You describe the quarrel as one of deciding how far to the left the party should be. More accurately the emphasis should be the other way round.

On the question of policy, the Right-wing element has long since showed that it is not interested in any progressive Socialism. Partly for constitutional reasons the party has never been very actively Socialist, nevertheless the platform is still Socialism.

Thus, to compare the Australian situation with that of British Labour is false. In Britain, it is, at most, a difference in the degree of application of the same policy, whereas in Australia it is a difference in fundamental policy.

You suggest that Dr. Evatt's personality and outlook has wrecked the party. But the Right-wing Roman Catholic element has made it perfectly clear that they are out to control the party, and that they will not use it in the interests of the country as a whole, nor in the interests of Labour as a whole, but specifically in the interests of the Roman Catholic Church.

Could any leader keep the party together under these circumstances? This very same element was critical of Mr. Chifley before his death, whose outlook on important subjects, such as Communism, was very similar to that of Dr. Evatt. In fact this latest development was inevitable, the only criticism being that Dr. Evatt compromised for so long.

Although the emergence of a separate Roman Catholic party, in fact if not in name, will put Australian Labour back by many years politically, the majority of supporters will welcome the possibility that it can again become the militant organisation that has fought and won so much for the country in the past.

Yours faithfully,
MILES E. WOHLERS.
London, W.14.

B.B.C. ELECTRICIANS

To the Editor of The Daily Telegraph

SIR—It is with much surprise I note that it is reported that the B.B.C. has recognised the Electrical Trades Union respecting staff negotiations.

Since this is the Communist-led union which was responsible for holding up the national Press for several weeks, shall we now soon be subjected to a cessation of radio and television services on the slightest pretext of some grievance?

Yours faithfully,
H. A. PARFREMENT.
Godalming, Surrey.

WHICH BUILDINGS?

To The Editor of The Daily Telegraph

SIR—A deep gulf in taste exists between the respective advocates of traditional and modern design for new buildings in London.

Those favouring traditional design have one advantage. Most people know what is meant by Palladian, Regency or other styles, and can refer to specific examples and comment upon them if so minded.

With modern styles there is no such label. The nearest description their admirers can coin is "a design in contemporary idiom," words which can mean almost anything.

We should know what we are talking about if someone would mention three or four contemporary buildings in London which can evoke real approval from an observer in the street. I say "from the street" purposely. Internal efficiency is not enough: it is the outside of buildings that goes to make up the glory of a city. Yours faithfully,
Brighton. JOHN F. CHAPMAN.

DOLLAR CHANGING

To the Editor of The Daily Telegraph

SIR—Besides its considerable civilian function, Prestwick is the main airfield in Britain for America's military air transport service traffic. When I was passing through it recently a number of American Servicemen were buying souvenirs and making other small purchases at shops in the terminal. Some of them were paying for those in American coin.

They were allowed six shillings to the dollar on coin, as opposed to seven shillings for a paper dollar. This caused some resentment, expressed in remarks like "The British are chiselling us again."

Two American officers with me agreed that, if it were explained by simple notices why banks always charge more for exchanging coin than note money, an apparent injustice would be explained. It would be particularly worth while for those Americans who came from beyond the Mississippi, where the silver dollar is as common as the paper one. Yours faithfully,
R. N. BRETT-SMITH.
Washington, D.C.

FOX AND CAT

To the Editor of The Daily Telegraph

SIR—Peterborough has referred to the cat-like characteristics of the fox. There is a further similarity in the tracks.

The fox, like the cat, places its hind feet in the prints of its forefeet. It is often difficult in snow to distinguish a fox's tracks from those of a cat, but from time to time a fox drops its tail and leaves a faint brushing mark. Close examination may also reveal the marks of a fox's claws, which, being of the dog family, it cannot withdraw like a cat. Yours faithfully,
A. E. R. GOULTY.
Bowdon, Ches.

DUKE OF EDINBURGH

The Duke of Edinburgh will attend the Regimental Dinner of the Welsh Guards, of which he is Colonel, at the Dorchester on June 9. On Wednesday, June 29, he will visit the 4th/5th Battalion, The Queen's Own Cameron Highlanders, of which regiment he is Colonel-in-Chief, at Stobbs Camp, Hawick, Roxburghshire.

On June 30 the Duke will re-name the new schooner of the Outward Bound Moray Sea School at Springfield Quay, Glasgow. He will attend the Amateur Athletic Association's Championships at the White City Stadium on Friday, July 15.

COLOUR CEREMONY

Princess Margaret is to present new Colours to the 1st Bn. Suffolk Regt. on May 23. The old Colours were presented to the 1st Bn. East Suffolk Regt. in 1849.

They are the last of the large 6ft Colours of British Infantry and are [illegible] mounted by spearheads. They were taken to Australia in 1854 and later to New Zealand where they went into battle for the last time during the Maori War of 1860-66. The Suffolks are now in Germany.

CHRISTENING

The second son of Mr. and Mrs. C. M. Williams was christened Richard Charles on Saturday at Hartley Wespall Parish Church. The Rev. A. T. Humphrey officiated and the godparents are Major T. Bailey, Mr. Thomas McNair Scott, the Hon. Mrs. W. Macauley and Miss Sonia Camalet.

LUNCHEONS

Constitutional Club

The Political Committee of the Constitutional Club yesterday entertained Sir Stephen Pierssené, General Director of the Conservative Central Office, to luncheon. Afterwards he spoke to a large gathering of members. Mr. R. F. Millard presided and the vote of thanks was given by Sir William Grimshaw.

Anglo-Brazilian Society

The Council of the Anglo-Brazilian Society gave a luncheon party yesterday at 2, Belgrave Square, in honour of Prof. Lucas Nogueira Garcez, until lately Governor of the State of São Paulo, and Senhora Garcez. Sir Donald Gainer, Chairman, presided.

DINNER

High Commissioner for Australia

The High Commissioner for Australia and Lady White gave a dinner party last night at Stoke Lodge, 45, Hyde Park Gate.

Those present were:

Air Chief Marshal Sir Donald and Lady Hardman, Air Vice Marshal Sir Robert and Lady George, Sir Keith Officer, Mr. and Mrs. McClure Smith and Miss Macdonald of Sleat.

SERVICE DINNER

Kut Garrison

The survivors of the Siege of Kut-al-Amara 1915-16 held their annual dinner at the Charing Cross Hotel on Saturday. Major-Gen. H. H. Rich presided and 86 of all ranks were present. The guest of honour was Sir Thomas White, High Commissioner for Australia.

IN MEMORIAM

Lt.-Col. T. W. C. Carthew, Q.C.

A memorial service for Lt.-Col. Thomas W. C. Carthew, Q.C., was held yesterday in the Temple Church. Canon J. D'E. Firth, Master of the Temple, officiated.

Many members of the Middle Temple and other well-known legal figures attended the service. Relatives present were:

Mrs. Carthew (widow); Miss A. Fanshawe (niece); and Brig. M. Carthew (cousin).

Latest Wills

	Net
BROTHERTON, H., Llandudno (gross £31,938)	£31,137
BUCKLE, Mrs. M. A., Shanklin (gross £21,358)	23,686
HALL, R., Helston, Cornwall, dental surgeon (gross £122,787), duty paid [illegible]	122,277
HAMMERSLEY, Mrs. L. M., Long Crendon, Bucks (gross [illegible])	[illegible]
HEWERDINE, H. W., Bournemouth (gross [illegible])	[illegible]
IRELAND, A. B., Altrincham, Ches. (gross £128,889), duty paid [illegible]	[illegible]
NAPTHEN, E., Penwortham, Lancs, solicitor (gross [illegible])	[illegible]

Court [and Social] Circular

BUCKINGHAM PALACE, May [...] The Duke of Edinburgh [was] present this evening at the [...] Anniversary Banquet of the [Insti]tute of Chartered Accountants [in] England and Wales at Guildhall.

Squadron Leader Beresford Ho[rsley] was in attendance.

ST. JAMES'S PALACE, May [...] The Duchess of Gloucester [was] present this evening at a Perf[orm]ance of "Upon This Rock" [in] Peterborough Cathedral to celeb[rate] the 13th Centenary of [the] Foundation of the Abbey.

Mrs. Cedric Holland was in attendance.

COPPINS, Iver, May [...] Princess Alexandra of [Kent], attended by the Lady Moyra Ha[mil]ton, this afternoon opened [the] Slough and District Arts Fes[tival] and Exhibition at the Slough C[om]munity Centre.

The Duchess of Kent saw ["An] Evening with Beatrice Lillie" at [the] Globe Theatre last night.

The Duchess of Kent will atte[nd a] Council Meeting of the Royal [Air] Force Benevolent Fund, of which [she] is President, at Portland Place [on] Monday. On May 17 she will [visit] Southend to name a new lifeboat [and] open the Carnival Estate for [Old] People, and on May 18 will [visit] Uxbridge to present the Charte[r of] Incorporation to Uxbridge Urban [Dis]trict Council.

Princess Alexandra of Kent will [visit] the Guide Dogs for the Blind Ce[ntre] at Leamington Spa on Saturday.

The Prime Minister of Northern [Ire]land, Viscount Brookeborough, since his return from Australia [and] New Zealand has been suffering [from] a recurrence of duodenal troub[le. He is] reported to be making satisfac[tory] progress. A further period of [rest], however, is necessary, and on med[ical] advice he will not undertake any [en]gagements during May. He expec[ts to] resume his public work next mont[h].

Air Chief Marshal Sir Arthur L[ong]more, vice-chairman of the Imp[erial] War Graves Commission, has [been] prevented by illness from acc[om]panying the Duke of Gloucester [dur]ing his tour of Commonwealth cemeteries in Italy.

Gen. Sir Brian Robertson, ch[air]man of the British Transport C[om]mission, who on Saturday was st[ruck] on the cheek by a polo ball, left [hos]pital yesterday for his Glouces[ter]shire home. He will be away f[rom] his office for a few days.

The opening of the new wing [of] Saffron Walden Training College [on] May 13 by Mr. R. A. Butler has b[een] postponed.

A memorial service for Lady Ann[aly], whose death took place yesterday, [will] be held on Friday at St. P[aul's], Knightsbridge, at 12.15 pm.

A memorial service for Vice-Ad[miral] Sir Forrester Rose will be held to[day] at St. Columba's, Church of Scotl[and], Pont Street, at noon.

TO-DAY'S BIRTHDAYS

Viscount Stansgate is 78 to-[day]; Air Vice-Marshal Sir Alfred Irede[ll] 76; Gen. Sir Eric de Burgh 74; [Lord] Howard of Glossop 70; Lt.-Gen. [Sir] Ralph Eastwood 65; Sir Alwyn C[...] 61; Sir Ewart Levy 58; Sir James Jejeebhoy 46; Earl of Ellesmere [...] and Lord Leslie 23.

FORTHCOMING MARRIAG[ES]

Mr. C. H. J. R. Joynson and M[iss] F. B. Preston

The engagement is announ[ced] between Lieut. James Ram[...] Joynson, The Loyal Regiment, el[der] son of Lt.-Col. and Mrs. N. H. Joynson, of Yatton Court, Bea[...] N. Devon, and Faith Beat[rice] (Dickory) Preston, daughter of [the] late Mr. D. E. Preston and [Mrs.] Preston, of Duddon Lodge, Tarpo[rley], Cheshire.

Mr. N. F. Rosier and Mrs. J. S[...]

A marriage has been arranged, [and] will take place quietly in June, [be]tween Norman Frank Rosier, of [the] Old Malt House, Wanborough, W[ilts], son of Mr. and Mrs. F. C. Rosier, [of] Newbury, Berks, and Joan, daug[hter] of Dr. Eric Wordley, M.C., and [Mrs.] Wordley, M.B.E., of Windrush, [St.] Mawes, Cornwall.

Mr. M. H. Denyer, R.M., and Mi[ss] J. J. Teakle

The engagement is announ[ced] between Michael, elder son of [Mr.] P. H. Denyer, of Cyprus, and Mrs. A. Denyer, of Hove, and Jenifer, daugh[ter] of Mr. and Mrs. H. O. Teakle, of N[...]bourne, Chichester.

Mr. A. Butler-Bowdon and Miss [...] Pollock

The engagement is announ[ced] between Anthony, younger son of Col. and Mrs. W. Butler-Bowdon, [of] Dapeland, Mayfield, Sussex, [and] Marion, only daughter of Mr. [and] Mrs. J. C. K. Pollock, of Newla[nds], Cape Town.

Mr. D. J. Cruickshank and Mis[s] A. K. O'Meara

The engagement is announ[ced] between Donald James, younger son [of] Professor and Mrs. J. C. Cruicksha[nk], 48, Lanchester Road, London, N.6, [and] Ann Kathryn, only daughter of [Mr.] and Mrs. T. B. O'Meara, 5, Cambri[dge] Mansions, Walton-on-Thames.

Mr. D. N. Pearson and Miss S[...] McIlwraith

The engagement is announc[ed] and the marriage will take pl[ace] shortly, between Derek Nigel Pears[on], son of Col. H. S. I. Pearson and st[ep]son of Mrs. Pearson, of Green H[...], Benson, Oxford, and Sheena [Mc]Ilwraith, only daughter of Mr. [and] Mrs. J. McIlwraith, of 7, Corstorph[ine] Hill Avenue, Edinburgh.

Mr. P. N. Burroughs and Miss J. K. W. Potter

The marriage arranged between [...] Peter Burroughs, R.A., and [Miss] Jenefer (Lin) Potter, of Westerha[m] [illegible] S.W., on June 25 at 2 p.m.

Mr. A. Peters and Miss S. R. Po[tts]

The engagement is announ[ced] between Alan, younger son of the l[ate] Mr. W. H. Peters, O.B.E., and of M[rs.] A. W. Durrant, of Eastbourne, [and] Susan Rutherford, daughter of [the] late Mr. M. J. Potts, M.B.E., and Mrs. M. A. Potts, of Alford, Lincs.

Mr. M. J. B. Price and Miss E. Le Gal-Robinson

The marriage will take place May 28 between Michael Jo[...] Bonamy Price, only son of Mr. a[nd] Mrs. W. de G. Price, of Guildfo[rd], and Edna Francis, only daughter [of] the late Mr. R. Hambro Le G[al-]Robinson and Mrs. Le Gal-Robins[on], of 11, Croft Gardens, Ruis[lip], Middlesex.

Mr. M. G. Porteous and Miss U. Flitch

The engagement is announc[ed] between Malcolm Geoffrey Porteo[us], B.A., only son of Mrs. E. Porteous, 52, Iddesleigh Road, Bournemou[th], and Ursula Mary, only daughter of [the] Rev. and Mrs. Flitch, of Highfie[ld], Harestock, Winchester, Hants.

Lieut. R. S. K. Seeley, R.N., an[d] Miss S. P. Laughton

The engagement is announc[ed] between Rodney Sidney Kingsley, o[nly] son of the late Mr. H. G. Seeley a[nd] of Mrs. Seeley, 51, Oxford Garde[ns], W.10, and Sylvia Patricia, young[er] daughter of Major F. E. Laught[on] and of Mrs. Laughton, 12, Thorn[...] Road, Mannamead, Plymouth.

Lieut. J. C. Q. Johnson, R.N., an[d] Miss B. U. Shelley

The engagement is announc[ed] between John Cecil Quarry, son [of] the late Mr. R. R. Johnson, of C[...] and Mrs. Johnson, St. Margaret's [...] and Bridget Ursula, daughter of [Mr.] and Mrs. R. Shelley, of Penang.

Mr. E. A. Hoare and Miss D. Hea[ley]

The engagement is announc[ed] between Edward Alexander (Pete[r]), only son of Mr. and Mrs. Percy Hoa[re], of Sanderstead, Surrey, and Dore[en], only daughter of Mr. and Mrs. B. Healey, of Bexhill-on-Sea, Sussex.

Mr. R. J. L. Cronchey and Miss K. M. Nye-Chart

The engagement is announc[ed] between Robert Julian Lawren[ce] Cronchey, of Longonot, Kenya, elde[r] son of the Rev. Dr. T. H. and M[rs.] Cronchey, of Moulton Vicarag[e], Spalding, Lincolnshire, and Kathle[en] Mary, daughter of Mr. and Mrs. [...] Nye-Chart, of Gilgil, Kenya, E. Afri[ca].

Mr. D. Hewerdine-Koop and Mis[s] M. E. Palfrey

The engagement is announc[ed] between David, son of the Rev. a[nd] Mrs. H. V. Koop, of Chedburgh R[ec]tory, Bury St. Edmunds, a[nd] Margaret Ellen, daughter of Mr. a[nd] Mrs. E. Palfrey, of 32, Manor P[...] Gardens, Edgware, Middlesex.

Mr. R. Gabriel and Miss E. N. [...]

[illegible]

ARTS COUNCIL AWARD

[illegible]

Engine driving is also a science

May 10 1955

Grand Bank, Edward Road W
Clevedon Somerset

Dear Sir

I have just read with pleasure your letter in today's *Daily Telegraph* on Engine Driving as an Art, and as a medical man would add as a Science.

For all those of us who are conscious of the trust for life and limb we place in the engine driver when we enter a railway carriage to journey by night or day you have well and truly expressed our appreciation. Politics have no place in such a concern. What he is paid should bear a just relation to his responsibility for human life and safety.

Yours truly Cecil W. Ewing

64 Morse Street
Swindon Wilts

May 12 1955

Dear Sir

Please accept this letter of mine in warm appreciation for a very lovely tribute to engine drivers in the *Daily Telegraph* of Tuesday May 10.

My husband is an engine driver but at the moment very very ill after a sudden collapse in January. Only on Monday last I accompanied him to Oxford for admission to the Churchill Hospital for observation and further treatment. He's a steady, conscientious fellow, having served with the GWR for 35 years.

In industry a worker received his wages after having given a few years should he become sick – then why not our engine crews?

Again many thanks

Yours truly Mary Wilkins (Mrs)

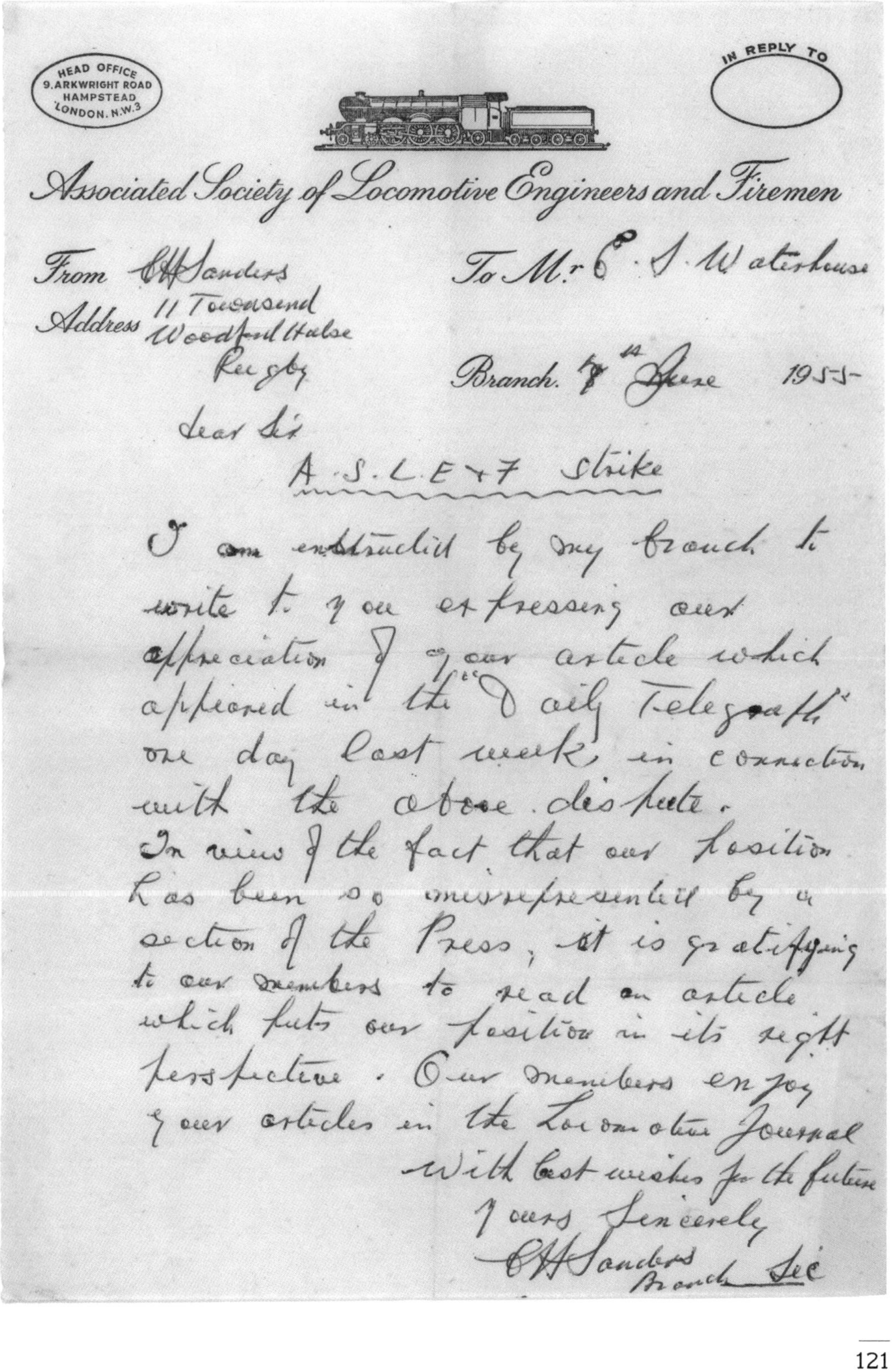
HEAD OFFICE
9, ARKWRIGHT ROAD
HAMPSTEAD
LONDON, N.W.3

IN REPLY TO

Associated Society of Locomotive Engineers and Firemen

From C H Sanders
Address 11 Townsend
Woodford Halse
Rugby

To Mr. C. S. Waterhouse

Branch. 14th June 1955

Dear Sir

A.S.L.E & F Strike

I am instructed by my branch to write to you expressing our appreciation of your article which appeared in the "Daily Telegraph" one day last week, in connection with the above dispute.
In view of the fact that our position has been so misrepresented by a section of the Press, it is gratifying to our members to read an article which puts our position in its right perspective. Our members enjoy your articles in the Locomotive Journal
With best wishes for the future
Yours Sincerely
C H Sanders
Branch Sec

Two glorious runs with LNER Pacifics

June 1955

The Flying Scotsman has become more sedate, leaving the fireworks to The Elizabethan. The winter allowance of 304 minutes for 268 miles to Newcastle, with a six-minute stop to change engines at Grantham, allows a 475-ton train time enough to recover from smaller delays and arrive punctually. I travelled behind A1 Pacific Scottish Union, Bros Widdowson and Harrison (Grantham). Though the engine was getting near to an overhaul, she steamed readily.

Starting slowly we took nine minutes to pass Finsbury Park, but then quickened to pass Hatfield, 15 miles on, in another 18 minutes. A bad signal check at Welwyn cost four minutes, but passing Hitchin to the stop at Grantham, including the 20 mph through Peterborough, we needed only 69½ minutes for 73½ miles and were in 4½ minutes early. Here a similar engine, King's Courier, looking smart after a recent visit to the shops, with Bros R Watson and Lindley of Grantham, took over. A pw slack before Retford cost two minutes, but we forged ahead with a clear road and during the last 20 miles the engine was eased to prevent too early arrival. Nonetheless, despite severe signal checks at Durham, we rolled into Newcastle four minutes early – an excellent run all the way. Congratulations to both crews.

The return was on the Tees-Tyne Pullman. True, it gets 33 minutes more than the famous Silver Jubilee had, but the load was 320 tons against loads from 220 to 278 tons. From Darlington the non-stop run is allowed 230 minutes for 232 miles. Bros Howard and Wilson of King's Cross had A4 Pacific Dominion of Canada, which was newly out of the shops and Darlington reached in only one more minute than Silver Jubilee was allowed.

Starting again we had our share of delays – pw at Raskelf, almost a dead stop by signal at Heyworth box, signals at Retford and Grantham, pws at St Neots and Woolmer Green. But nothing could hold us down. With the liveliness of an electric motor we got away again after each check and galloped ahead, finally running in five minutes early after a run comparable with Silver Jubilee itself when the extra load and severe delays are taken into consideration, to the credit of enginemen and engine alike.

If the days of steam are slowly ending, the end, with runs like the two described here, will be glorious!

Good and bad to and from Crewe

February 1956

My duties take me frequently to North Wales and I travel to Crewe on the Midday Scot, with loads varying from 400 to 465 tons. The 158 miles are timed in 161 minutes, including a four-minute stop at Rugby. So far every trip has been excellent.

The last was behind City of Glasgow, one of the engines involved in the up Harrow disaster. Bros Parrish and Ashmore, of Crewe, were in charge and, as the engine had been riding rather rough, Inspector Morgan travelled with them. The load was 404 tons.

We drew away smartly and passed Tring summit, 31¾ miles, in the booked 35 minutes. The first 60 miles took a shade under 57 minutes and then we met severe relaying and entered Rugby three minutes late, saved a minute station time, then met further pws and were four minutes late passing Nuneaton, but by reeling off the 36½ miles in exactly 30 minutes we passed Stafford two minutes early. Further delay wiped this out, but a storming finish down Whitmore bank saw us into Crewe one minute early.

True, we were a little below the maximum load of 453 tons (sometimes exceeded), but the way the engine was handled suggested that it would have been able to take that load just as easily. I have a high opinion of these engines and of the Crewe North link who work them.

The return is a less happy story. I join the Irish Mail from Holyhead, worked with a Scot or a Britannia, and have not yet booked a punctual arrival in London. Holyhead men work to Crewe, and we are pretty near our time if we are not stopped outside. The train almost invariably exceeds the seven-minute station allowance, although the engine works through.

On my last trip we left seven minutes late with Scot class 46147. Bros Bailey and Jones, of Crewe, had 460 tons behind. Two relaying checks and signals caused us to take 49 minutes to pass Rugeley, 33¾ miles, and then followed distress signals of shortage of steam and we stopped at Rugby to get assistance, having taken 100 minutes for the 75½ miles. A Stanier Black 5 diverted from carriage shunting, we started off and finally got in 38 minutes late.

It is difficult to say why this train is so consistently a poor runner. I am sure that if it were possible to change engines at Crewe, and give it a Duchess, it would return to the right paths. I have great respect for the Scots, but 460 tons is a bit too much for them unless one gets an engine in better condition than most are now. I could find nothing to criticise in the way the train was worked

Five readers' questions answered

Responses to questions asked by readers over the past month or two. These are my opinions only, and I may be wrong.

July 1956

Where is the best firing?

On the Western. Much "blind" coal is used and it must be broken up and incandescent before the required heat can be got from it. Western men have to fire little and often, to break up the coal and keep the flap down – even to the extent at times of closing it after every shovelful.

Which is the best engine?

None. It all depends on best for what. If I had to choose one engine for all classes of work, I should select a Stanier Black 5, or an Eastern B1. Possibly even a Western Grange, for I was once amazed at the ability of 6810 running non-stop London to Newport with 375 tons in a net time of 138 minutes for 133½ miles. Or, in good form, I'd even back a Southern Arthur to do anything reasonable at all kinds of jobs.

Is the 4-4-0 type done for?

No. The Southern Schools class still do splendid work. But no more will be built, I imagine. The LMS compounds were excellent in their day, and so were the George Vs and the Great Central Directors.

Do you favour full regulator and early cut-off as normal practice?

Yes, but I admit not as surely as I once did. It applies mostly to express work. Most mixed traffic engines need 20 per cent at least, and I have seen on the footplate good results from small regulator and longer cut-off. In theory, to use the expansive power of steam is certainly correct, but one has to find out what suits each type best.

Why are there individual differences between similar engines?

A wag might say because of their cranks and eccentrics! But, seriously, I don't know. Anyhow, no engineman will deny these differences, even if there is no satisfactory explanation of them. Even after a visit to the shops good and bad engines seem to retain their character from beginning to end of their career. Sometimes a good one loses its virtues, but it is rarely that a poor one shows much improvement.

EDITOR'S FOOTNOTE

ESW wasn't short on opinions, but when writing for enginemen he was in restrained mode. Perhaps the most interesting questions are about practical issues met every day.

Dugald Drummond and his engines

October 1956

Just on 60 years ago South Western enginemen were curious to see what the Drummond bogie tanks were like. They were the first, and as it turned the best, that this hard-headed, hard-swearing old Scot designed for the LSWR.

For 14 years they continued to be built with only minor changes. They have 18½in x 24in cylinders, 5ft 7in driving wheels, and tractive effort of just under 20,000lb with 175lb pressure. Of the 0-4-4 type, they were not suitable for fast work but oddly enough were intended for the Plymouth-Exeter section mainline traffic. They were soon taken off that, and finally became responsible for London suburban traffic. When displaced by electricity they migrated to country branches. About 20 of them may be seen taking stock in and out of Waterloo any day.

In all 105 were built and 103 are left, all as they were. No 126 was rebuilt with superheater, but though she was "two coaches better" as a result the extra weight caused overheated boxes and she was broken up. No 672 took a dive into the bowels of the earth, falling down a service shaft into the Waterloo and City tube. The rest remain to this day, a wonderful record indeed. Drummond followed with the 0-6-0 goods, the Black Motors, 30 engines all still at work.

The other famous class, the Greyhounds or T9 4-4-0 type, did the bulk of the mainline work for many years with great success, though those with Drummond's cross-tube fireboxes were later converted to the ordinary type. Most of them remain and are found on cross-country routes as a rule, though you may see them still on the fiercely-graded Devonshire sections. Drummond was not afraid to experiment, though the 4-2-2 express engines which had four cylinders and uncoupled driving wheels, the "slipper-blocks", were cordially disliked.

Drummond's 4-6-0 express types were moderately successful, the last batch, the Paddleboxes, being the best, but all have gone now. So has his little inspection car with a 4-2-4 tank engine, and his 15 rail-motors designed for branch work. His later 4-4-0 expresses, 463 class, have not outlived the T9s, as all have gone, and so have the 4-4-0 mixed traffic engines known as Hoppers, small and big hoppers, in all 80, which again were not popular.

Personally, Drummond was kind to me when I saw him as a lad and asked for a job, though I was not able to take his offer, but he was too sharp-tempered to be liked by his men; yet all admitted that he knew his job.

A fireman looks back to life in the nineteenth century

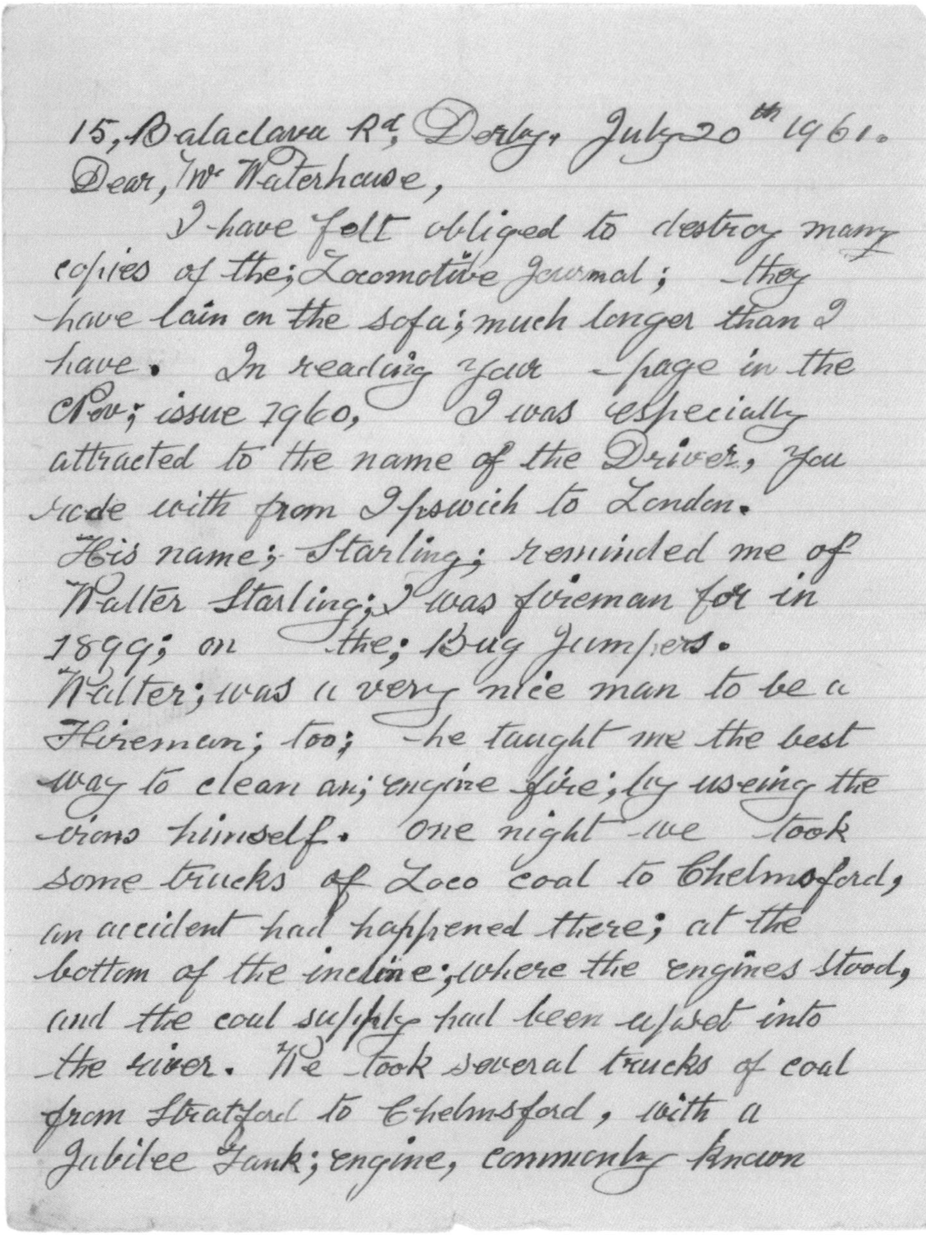

15, Balaclava Rd, Derby, July 20th 1961.
Dear, Mr Waterhouse,
I have felt obliged to destroy many
copies of the; Locomotive Journal; they
have lain on the sofa; much longer than I
have. In reading your page in the
Nov; issue 1960, I was especially
attracted to the name of the Driver, you
rode with from Ipswich to London.
His name;- Starling; reminded me of
Walter Starling; I was fireman for in
1899; on the; Bug Jumpers.
Walter; was a very nice man to be a
Fireman; too; he taught me the best
way to clean an; engine fire; by useing the
irons himself. One night we took
some trucks of Loco coal to Chelmsford,
an accident had happened there; at the
bottom of the incline; where the engines stood,
and the coal supply had been upset into
the river. We took several trucks of coal
from Stratford to Chelmsford, with a
Jubilee Tank; engine, commonly known

as a; Bug Jumper, class of engine.

The gradient of Brentwood Bank; had'nt been reduced then; but we managed to reach the summit; at seven arches, in about the usual time allowed. We refilled the water tanks at Chadwell Heath; and Shenfield; on the journey. I have no recollection of hearing the name of Starling; in reference to a person, since I was Walters, fireman. Some of the mates called him; sparrow; it did'nt affect him. Now I'm wondering if the Driver of John Bunyan; was a relation; a descendant of Walter. I'm speaking of the time; before the the; South African War.

Yours fraternally
Robert; Bob Scott.
EX-FIREMAN 173.

P.S.
I'm thinking the; two "Starlings", were perhaps related. R.S.

End of a branch line in West Sussex

November 1963

With the remorselessness of Judge Jeffreys' Bloody Assizes the branch lines are being marked down, and it is sad to hear that the Horsham-Brighton branch is among them. It runs through the heart of West Sussex, through meadows and the Downs, to join the Brighton-Portsmouth line by the river at Shoreham where, as at Brighton, it connects with the whole South Coast and at Horsham with Waterloo, London Bridge and Victoria.

There are 17 trains each way on weekdays and 11 on Sundays. Most trains make 11 stops and the timing is just about an hour. That does not sound like a direct line which is past revival.

For 102 years it has served a useful purpose. When I first knew it the little LBSCR tanks of the D1 type did much of the work, and later the 0-6-2 class E4. After the Southern Railway came into being the Drummond M7 type took over and were seen until the last few years, when the BR 80 type arrived.

Tank engines have been predominant on the line from the beginning. I was travelling on this line lately and found that a number of ex-LMS 2-6-2 type tanks of Ivatt's design of 1946 were at work there. They have 5ft 7in drivers, 16in x 24in cylinders, 200lb steam pressure, taper boilers, Walschaerts gear and PV, with TE of 18,510lb and seem well-liked.

The trains are light, three bogies, and keep excellent time. Two of our Horsham members, Bros Hill and Tanner, worked my train to Shoreham with 41312, and two of our Brighton members, Bros Colley and Pannett, brought me back with 41327. In all my travels this year it is only the second time when I have been steam-hauled, more's the pity!

To close lines without proper and equally convenient other transport is to treat railways like multiple stores – pay or close. Instead of throwing passengers and freight onto roads which, the experts forecast, will be trying to carry 13 million vehicles by 1970, sanity surely calls for the Labour Party's proposal to consider road and rail policy together.

The Tories, the Big Businessmen's party, might ask themselves which is worse, the deficit lost on the national necessity of rail transport, or millions gone down the drain in "deterrents" that won't function and are quietly scrapped like Blue Streak. But, as John Bright said once, "The Tories are always a stupid party."

London-Manchester in two hours?

January 1964

It has been said recently by a railway traffic manager that the future will see 150 miles per hour, with Manchester in two hours and Glasgow in five. In the former case it will be electric power, but that is unlikely to apply to the latter which will be, for a long time yet, reached by diesels.

By the way, it is just over 50 years since Rudolf Diesel, the inventor, disappeared from the Harwich boat crossing to England. Some say accident, some whisper murder, but none knows.

Attaining 150 mph is not impossible. The French claim that their electric trains have done 200 mph, but it is said that the result was very expensive for the permanent way, and if we, who so far have speed limits of 80-90 mph for diesels, are to make these the average speed, it will need some new designs.

The line from Euston to Manchester is suitable for speed. Once away from Willesden high speed could be maintained, with a slight reduction for the curve at Weedon, to Rugby and then to Stafford. From Whitmore to Crewe the steepest grade on the journey, 1 in 177, is met. On this falling grade, on a trial run, Pacific 6220, Driver T.J.Clarke and Fireman J.Lewis of Crewe, touched 114 mph. The average speed for the run from Euston was just under 80 mph for the 270 ton train.

Manchester is some 31 miles from Crewe and the slowing down at Crewe and the necessarily slow approach to Manchester would mean that at best 25 minutes would be needed, so that to reach Manchester in two hours it would be needful to pass Crewe, 158 miles from Euston, in 95 minutes at around 100 mph, but with the inevitable reduction of speed for signals, junctions, curves and pw slacks, accurate timekeeping would require a higher than average recovery time. I do not say it would be impossible, but it would be expensive in track maintenance, fuel, and wear and tear to locomotives and coaches.

Does the average passenger greatly care if, let us say, a 300-mile journey takes three or three-and-a-quarter hours? Is not a comfortable journey, punctual, at reasonable cost, more likely to attract him than very high speed? Perhaps some of my readers will one day be driving these 150-milers. I should be interested to hear what they think of their possibility and value.

A steam glossary

The working life of ESW's readers revolved around steam locomotion, which had a vocabulary of its own, so terminology in his columns carries few explanations. This glossary attempts to define words and phrases commonly used in the columns. For more on technical terms (marked with an asterisk) visit Wikipedia or railway enthusiast websites.

Axle box wedge* – protective steel wedge between the axle box and the horn block to maintain clearances

Blind coal – coal of uncertain quality, difficult to fire

Barracks – railway staff lodging

Blower* – steam jet to draw fire

Bogie* – shorthand for leading-wheel truck, but also slang reference to the number of carriages in a train

Booster* – additional power source for starting heavy train or maintaining low speeds in tough conditions

Brick arch* – firebox component

Check – speed restriction (see also pw, pws, pwr, slack)

Coal-dodgers – those involved in racketeering over the use of coal in the early 1900s

Compound* – engines having multiple high and low pressure uses of steam

Coupled – engines with linked driving wheels, usually two on each side

Cut-off* – control of steam consumption and performance through locomotive valve gear

Distant/distants – signals with advance cautionary warning of stop signals and track restrictions

Down – usually services from London

Eccentric* – rod operating steam valve gear

Engine slip – wheel slip on starting or gradients

Flap – firebox hole cover

Foreign – engine not operated from home shed. Sometimes refers to locomen from other depots or 'foreigners'

Frame* – substructure or chassis of locomotive

Heating surface* – combined total surface of firebox and tubes

Home signal – A stop signal, in advance of a signal box

Hooker – slang for engine attached as additional power for heavy or difficult loads, more properly called an assisting engine. The practice was also called double heading

Horn block* – part of an axle-box frame

HP chest* – high-pressure steam chest in compound loco

Link – team of enginemen and guards operating regular services or rosters. Typically a link would comprise 12 men working designated shifts

Loading-gauge* – measurements for clearance of tunnels, bridges etc

Lodging – practice of overnighting on certain turns at lodging houses or company barracks

Loose-coupled – a freight train without continuous brakes, the only brakes being on the loco and in the guard's brake van

Mileage runs – long-distance turns on which crews got a mileage bonus

Notching up – action to limit steam pressure in cylinder (see cut-off)

On the shovel – firing

Passed cleaner – first qualification to become a fireman

Pw, pws, pwr – permanent way speed restrictions

Pony truck* – single-axle leading bogie enabling loco to take curves more easily

Recovery – time regained against schedule

Road – permanent way

Running table – log of intermediate journey times/events

Sanders – dispensers of sand onto rails to improve traction, usually steam-powered

Shed – local base where locos were stabled and maintained

Shop block – instruments as set in the works

Singles – engines powered by a single axle driving wheel

Slack – speed check

Slack coal – poor quality coal

Slip* – practice of uncoupling coaches without stopping

Speed orders – speed restrictions

Spinner – slang for engine with single axle driving wheel

Station pilot/banker – loco kept at stations for relief work or shunting

Sticks – slang for signals (also 'pegs', 'boards' and 'sigs')

Stroke/bore* – Measurements of cylinder dimensions

Superheating* – refinement for improved steam performance

Tank – tank engine

Tare* – weight measurement of trailing load, not including loco

Tractive effort (TE)*- engine pulling power stated in lbs based on pressure of wheel on rail

Turns – rota or roster work

Up – usually services to London

Wheel pick up – normally when loco wheels lock or slide on the rail

Endword

ESW was of his generation. He celebrated the Pacifics of Stanier and Gresley as the finest British examples of railway steam power – by its nature noisy, dirty, demanding, and fossil fuel greedy. He loved the countryside but was no environmentalist. His father had made his rural doctor's rounds on horseback; he drove petrol-engined motorcycles and cars throughout his adult life, ending up owning a Mini.

He accepted that railways of the future would be electric but hated diesels, arguing without success the case for retaining steam until all-electric lines were opened. He held that diesels were more costly to build and maintain, more polluting than steam, and left dependence on oil-producing nations.

He believed that investment in a modern railway system was at least as important as the emerging motorway network. He deplored the Beeching cuts which decimated local services. But he was not an advocate of speed at any price.

His February 1941 column, looking from the depths of Blitzed Britain at the prospect of postwar state ownership, set out a management model in which employee and passenger representatives were given a powerful voice in the way railways were run. How much better might today's scene seem if this approach had been adopted and fostered.

He valued his many friendships with those who fired and drove the steam era. In his view, the long years spent graduating from cleaner to fireman to driver around the harsh footplate environment were as testing and worthwhile as any professional career.

In the issue of *Locomotive Journal* marking his 1964 death, the chairman of ASLEF's London Bridge branch is quoted as saying: "Professor Waterhouse thought more of Locomotivemen than many of them think of themselves."

Amen to that

Robert Waterhouse

Opposite top: ESW with wife Edie and son John, South London, 1916
Bottom: ESW with grandson Robert, North London, 1962

TBW639

Appendix

Above and cover:
Great Eastern
B12 on mainline
service near
Northampton,
1954

Left and overleaf:
Taking water at
Trafford Park shed,
Manchester, 1960s

A photographer's view of postwar railways: Steam, smoke and steel on steel

Denis Thorpe

In 1954, out of the corner of my eye, I saw a massive piece of metal approaching at speed. It was belching steam and smoke and the noise as it got near was of a great clanking of steel on steel.

I was standing near a mainline railway track in Northamptonshire where I had some forgotten assignment for the evening paper. I had my own preferred pre-war 35mm Contax camera with me and felt compelled to record that fraction of a second of amazing power. I pressed the shutter – one frame captured, no motor-drives available in those far-off days.

I wish I could talk about the number and class of this wonderful locomotive but I never acquired that ability. I remember that my brother and a cousin would visit one of the two stations in Mansfield and get excited about the numbers on the locomotives which passed through. I would only be interested in the power and the spirit of these incredible machines that breathed a sort of near-human life.

I was born in 1932 in a row of ten small cottages which stood up close to a main railway line. My mother told me later in life that as a toddler I was found standing near the track and saved by the watchful eye of a signalman.

When I became a journalist I saw these beautiful giants resting at stations and sometimes I recorded how they fitted into the landscape. Often on occasional journeys I would pull down the leather window straps and lean precariously out with my camera shutter on a slow setting so that I could capture a sense of speed.

Sadly, I had other work to cover and now regret that I did not pay enough attention to a subject which played such an important part in our transport system.

In the mid-1960s my friend Harold Pendlebury, who was chief reporter on a Manchester daily, persuaded me to visit the Trafford Park railway yards and I watched as crews took on water from giant pumps. Harold's father had been an engineer at Horwich locomotive works. Harold said it was urgent – we must record everything because it will soon be ancient history.

I must have blinked and then looked away – it was already too late.

Below: Midland 1528 class at Mansfield shed with the chimney of a former brickworks, 1950s

Above: Stanier 2-6-4 tank at Northampton station, 1950s

Overleaf: The surrounds of Mansfield Central station, which offered services to Nottingham until closure in 1956

THELSON MOTOR